Collective Modes in Inhomogeneous Plasma

Plasma Physics Series

Series Editors:

Professor Peter Stott, CEA Cadarache, France
Professor Hans Wilhelmsson, Chalmers University of Technology, Sweden

Other books in the series

An Introduction to Alfvén Waves
R Cross

Transport and Structural Formation in Plasmas
K Itoh, S-I Itoh and A Fukuyama

Tokamak Plasma: a Complex Physical System
B B Kadomtsev

Electromagnetic Instabilities in Inhomogeneous Plasma
A B Mikhailovskii

Instabilities in a Confined Plasma
A B Mikhailovskii

Physics of Intense Beams in Plasma
M V Nezlin

Forthcoming titles in the series in 2000

Plasma Physics via Computer Simulation, 2nd Edition
C K Birdsall and A B Langdon

The Theory of Photon Acceleration
J T Mendonça

Nonlinear Instabilities in Plasmas and Hydrodynamics
S S Moiseev, V G Pungin, and V N Oraevsky

Laser-Aided Diagnostics of Gases and Plasmas
K Muraoka and M Maeda

Inertial Confinement Fusion
S Pfalzner

Introduction to Dusty Plasma Physics
P K Shukla and N Rao

The Plasma Boundary of Magnetic Fusion Devices
P Stangeby

Plasma Physics Series

Collective Modes in Inhomogeneous Plasma

Kinetic and Advanced Fluid Theory

Jan Weiland

Chalmers University of Technology, Göteborg, Sweden
Euratom–NFR Association

Institute of Physics Publishing
Bristol and Philadelphia

British Library Cataloguing-in-Publication Data

A catalogue record for this book is available from the British Library.

ISBN 0 7503 0589 4 hbk

Library of Congress Cataloging-in-Publication Data are available

Production Editor: Simon Harris
Production Control: Sarah Plenty and Jenny Troyano
Commissioning Editor: Michael Taylor
Editorial Assistant: Victoria Le Billon
Cover Design: Jeremy Stephens
Marketing Executive: Colin Fenton

Published by Institute of Physics Publishing, wholly owned by The Institute of Physics, London

Institute of Physics Publishing, Dirac House, Temple Back, Bristol BS1 6BE, UK

US Office: Institute of Physics Publishing, The Public Ledger Building, Suite 1035, 150 South Independence Mall West, Philadelphia, PA 19106, USA

Typeset in $\TeX$ using the IOP Bookmaker Macros
Printed in the UK by J W Arrowsmith Ltd, Bristol

Contents

Preface

This book presents the collective drift and MHD-type modes in inhomogeneous plasmas from the point of view of both two-fluid and kinetic theory. It is based on a lecture series given at Chalmers University of Technology, Göteborg, Sweden on 'Low frequency modes associated with drift motions in inhomogeneous plasmas'. The level is undergraduate to graduate. A basic knowledge of electrodynamics and continuum mechanics is necessary and an elementary course in plasma physics is a desirable background for the student. The author is grateful to H Nordman, A Jarmén, P Andersson, J P Mondt, H Wilhelmsson, C S Liu, A Zagorodny, V P Pavlenko, A Rogister, R Singh and S C Guo for many enlightening discussions and to G Bateman, J Kinsey, A Redd and A H Kritz for their great efforts in predictive transport simulations in testing different transport models; it is gratifying for us that they have adopted our model as the most precise. Thanks are also due to B Jhowry, H G Gustavsson and M Hansen for using the scientific LaTeX word processing package to produce the manuscript and for help with the layout, and to H G Gustavsson, P Strand and J Anderson for help with proofreading. The American Institute of Physics and the IAEA are acknowledged for giving permission to reproduce figures 5.4, 5.12 and 5.13. Finally, I extend my gratitude to my family, Wivan, Henrik and Helena, for their continuous encouragement and support.

Jan Weiland

Chapter 1

Introduction

1.1 Collective Perturbations in Bounded Plasmas

For a plasma to be completely stable it must be in thermodynamic equilibrium. This is only the case if the plasma has a Maxwellian velocity distribution and is homogeneous in space. This means that a confined plasma will always be in a non-equilibrium state with different kinds of free energy available to drive instabilities. Already, Coulomb collisions will drive transport in an inhomogeneous plasma by scattering particles from one gyro-orbit to another. In a turbulent plasma the collective (turbulent) field will have a similar influence to the microscopic field in Coulomb collisions. There is, however, a phase difference required between density or temperature perturbation and electric field perturbation to obtain a net transport. For turbulence driven by linear instabilities, this phase difference is caused by the linear growth. The linear growth thus acts as a source of the turbulence at the same time as giving a correlation between density or temperature perturbation and the driving electric field necessary for transport. We can directly see the connection here between a deviation from thermodynamic equilibrium, linear instability and transport. This scenario, in general, leads to transport coefficients that depend on the gradients of density and temperature, thus leading to nonlinear diffusion or transport equations. As seen in chapter 7, the transport itself can also generate the phase difference necessary for its own generation. This can, however, often be seen as a result of a nonlinear saturation due to transport so that the turbulent phase shift equals the linear phase shift.

The reason for studying low frequency modes is that these generally give the largest transport. While the linear growth rate leads to an irreversible transport, the real eigenfrequency corresponds to a periodic reversible behaviour that reduces transport.

In the present work we shall consider both macroscopic magnetohydrodynamic (MHD) type modes and small scale drift-type modes. Since we need the more detailed two fluid or kinetic descriptions for the drift-type modes we shall also use these for MHD-type modes. This allows us to see the connections and make the transition between MHD and drift-type modes.

We shall also study the applicability of the two fluid approach by comparing it with the kinetic approach. This is only done here in the linear case. This comparison is also expected to give similar trends nonlinearly for the finite Larmor radius effects. For wave particle resonances, however, the situation may be very different since the nonlinear wave–particle resonance effects have a tendency to counteract the linear ones. Here we expect the sources in velocity space to play a crucial role. We may compare this with the situation in real space where a background gradient is necessary for transport and a background gradient on a long time scale requires a source.

In chapter 5 we study low frequency modes in a more realistic geometry and derive an advanced reactive fluid model for transport in toroidal systems.

Since the magnetic field confines a plasma in only two dimensions, the method of treating the problem with the third dimension is obviously very important. In a tokamak the toroidal curvature represents the third dimension. This means that the toroidal curvature is fundamental for the confinement. Its main obvious consequences are the presence of curvature driven modes and trapped particles. Since the curvature is driving instabilities only on the outside of the torus, curvature also leads to eigenmodes that are trapped on the outside. These are generally called ballooning modes. We note, however, that the term 'ballooning mode' was originally introduced for the MHD ballooning mode and this meaning is sometimes still assumed to be understood.

Although the effects of toroidicity mentioned above have been known and studied for a long time, it is only quite recently that strong efforts have been made to include the toroidal effects fully in calculations of tokamak transport. The main assumption to be removed from previous calculations of transport is that the diamagnetic drift, due to the pressure gradient, dominates over the magnetic drift, which is due to toroidal curvature and the closely related radial variation of the magnitude of the magnetic field. When this assumption is removed, a completely new regime of transport is introduced. This regime usually applies within 80% of the small radius of a tokamak. For shots with highly peaked pressure profiles, such as TFTR supershots, this regime is somewhat smaller, while for shots with flat pressure profile, such as the usual H-mode, it is larger. In the new regime, transport coefficients tend to grow with small radius, which is in agreement with experiments, and which previously was a main problem for drift wave transport models. Also, in the new regime the mode frequency is comparable to the magnetic drift frequency and this causes a major problem for fluid models. On the other hand, nonlinear gyrokinetic simulations are still far from being useful as transport codes on the transport time scale although much progress has been made over the last few years. For transport code simulations we are thus left with fluid or linearized kinetic models. This fact has recently led to the development of advanced fluid models that try to treat wave–particle resonances by a suitable closure scheme. These models are generally of a higher order, multiple pole type, since they include the full energy equation with its time dependence, thereby being able to make a continuous transition from adiabatic to isothermal states. The difference is that while some models retain the linear wave–particle resonances, as taken from linear kinetic theory, others, like the one described in chapter 5, assume that nonlinearities in velocity space effectively eliminate fluid moments of a higher order than those treated self-consistently. This leads to a reactive fluid model.

The significance of the ratio of magnetic to diamagnetic drift as the main toroidal effect for transport is shown by the fact that it is the largest one (this ratio goes to infinity towards the axis), and that it enters dynamically through the pressure gradient. These dynamics are important in transitions between different confinement states. It is also important to notice that a new stability regime is introduced for large values of the ratio between magnetic and diamagnetic drifts. In this region, the density scale length drops out and stability is determined by the ratio of temperature and magnetic field scale lengths.

1.2 Confinement in Fusion Devices

Traditionally, the theory of collective modes in magnetic confinement systems is divided into stability and transport. Stability then means stability of large scale magnetohydrodynamic (MHD) modes, which have to be stable to make a positive energy balance in a magnetized plasma possible. Transport is caused by small scale instabilities, which are very hard to control, but which have consequences that we may be able to live with, although we prefer to reduce them as much as possible. The transport due to collective modes is generally termed 'anomalous' since it is usually much larger than transport caused by the microscopic fields due to the finite extent of particles.

As will be shown later, the most dangerous modes for transport are those with small real eigenfrequency. Modes with a high real eigenfrequency mainly tend to make particles and energy oscillate in a coherent manner. Thus the main reason for our interest in low frequency modes in magnetized plasmas is the anomalous transport caused by these modes. One simple description of such transport will be given in section 3.2.8. Here we shall aim at a popular introduction, which will motivate the study of collective modes in magnetic confinement systems.

The quantity of main concern for the energy balance in fusion reactors is the parameter $n\tau_E$ where τ_E is the energy confinement time [1.1–1.3]. Since the fusion cross-section is almost a square function of the ion temperature in the regime 10 keV $\leq T_i \leq$ 40 keV, a measure of performance is the fusion product $n\tau_E T$. The experimental achievements here have improved greatly since the beginning of magnetic fusion research, and the development of the parameter $n\tau_E T$ as a function of time is shown in figure 1.1.

This improvement has been achieved partly by building larger machines and partly by optimizing the magnetic field geometry. The necessary condition

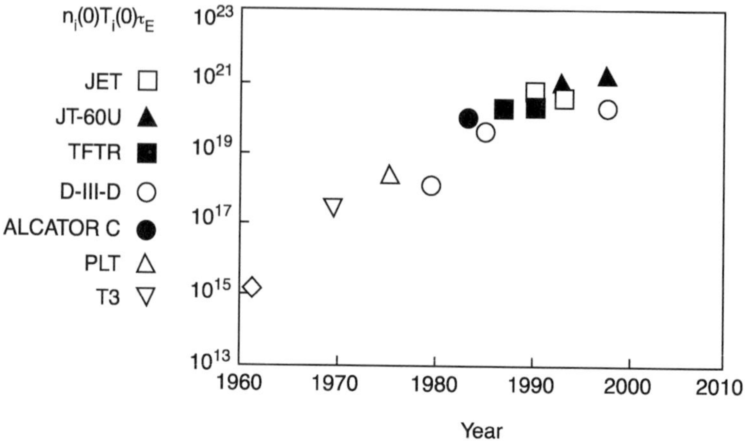

Figure 1.1. The development of achieved fusion product over the years.

to be fulfilled by $n\tau_E$ for nuclear fusion is the Lawson criterion [1.1]:

$$n\tau_E \gtrapprox \frac{3n^2T(1-\eta_t\eta_h)}{P_\alpha + \eta_t\eta_h P_n - (1-\eta_t\eta_h)P_r} \tag{1.1}$$

where P_α, P_n and P_r are the alpha particle, neutron and radiation powers, and η_t and η_h are the thermal and heating efficiencies. Since P_α, P_n and P_r are all proportional to N^2, equation (1.1) becomes almost independent of n. For a temperature of about 10 keV it takes the form

$$n\tau_E \gtrapprox 10^{20}\ \mathrm{m}^{-3}\ \mathrm{s}. \tag{1.2}$$

This is the condition for power *breakeven*. Since in this regime the fusion cross-section scales roughly as T^2, it is convenient to introduce the fusion product $n_i\tau_E T_i$. For ignition (only alpha particle heating) we get the condition

$$n_i T_i \tau_E > 5\times 10^{21}\ \mathrm{m}^{-3}\ \mathrm{keV\ s}. \tag{1.3}$$

Since experiments show that equation (1.3) cannot be fulfilled when MHD ballooning modes are unstable, i.e., when condition (3.29) is not fulfilled, we also obtain the condition

$$\tau_E > \frac{Rq^2}{a}\frac{10^{21}}{B^2/2\mu_0}\ \mathrm{s} \tag{1.4}$$

with $B^2/2\mu_0$ expressed in keV m^{-3}, which is now a condition on τ_E alone. Here a is the minor radius, R is the major radius, q is the safety factor (given by the ratio of toroidal angle to poloidal angle variations as we move along a field line, chapter 5) and B is the toroidal magnetic field. As an example this limit takes the value 4 s for JET with $B = 3.5$ T and $I = 4.8$ MA. The β limit (3.29) is due to MHD ballooning modes. When we also include the stability limit due to kink modes (chapters 3 and 5), the maximum average beta is given by the Troyon limit [1.4]:

$$\langle\beta\rangle \quad < \quad g\frac{I}{aB}\%. \tag{1.5}$$

Here I is the plasma current, and g is a numerical factor between 2.8 and 4.4 which depends on the elongation of the cross-section (ellipticity). When the plasma β is in the MHD stable regime, the confinement is on good grounds believed to be limited by turbulent transport. This transport has, in a rather wide regime for ohmically heated plasmas, been observed to follow the so-called Alcator scaling [1.5]:

$$\tau_E \approx 3.8 \times 10^{-21} na^2. \tag{1.6}$$

When this scaling is generalized to take into account the dependences on R and a it is called the neo-Alcator scaling.

This scaling has recently been recovered theoretically as due to the dissipative trapped electron drift mode [1.6] or the microtearing mode [1.7]. These modes are both driven by temperature gradients and the density dependence comes from a dependence on resistivity. A further discussion of the modes is contained in chapter 5. When the density reaches a high enough value the Alcator scaling is saturated and a region where τ_E is almost independent of n enters. The energy transport in this region is believed to be due to a drift wave driven by ion compressibility effects in combination with ion temperature gradients. This is the η_i mode [1.8] ($\eta = \mathrm{d}\ln T/\mathrm{d}\ln n$).

As it turns out, both the trapped electron mode and the η_i mode are, in the experiments, typically not far from marginal stability [1.9] in the so-called confinement region of the plasma. This is an indication that these modes actually govern the temperature profiles, giving rise to the so called *profile resilience* [1.10]. This is a typical feature observed in tokamak plasmas, where the temperature profiles are virtually independent of the power deposition profile by neutral beam or radio frequency heating. A close relation between modes driven by temperature gradients and energy transport is also expected from thermodynamic points of view, since a temperature gradient means a deviation from thermodynamic equilibrium, and since an energy transport would tend to equilibrate the system. In connection with auxiliary (non-ohmic) heating, a degradation in confinement (L mode) has been observed. In 1982, a new type of confinement mode, the H mode, was discovered on the ASDEX tokamak in Garching [1.11]. In this regime the confinement time is a factor of 2–3 larger than in the L mode. The confinement time does, however, degrade with power in the H mode as well. The transport research has, over the years, been conducted both by empirical and first principles methods. Empirically, one has derived scalings of confinement time with various characteristic parameters of the experiments. A very fruitful theoretical approach is to derive constraints on these scalings for consistency with the basic physics description [1.12, 1.13]. A general expression for the thermal conductivity can be expressed in terms of the Bohm diffusion coefficient $D_B = T/eB$ as

$$\chi = D_B \left(\frac{\rho}{a}\right)^{\alpha} f\left(\varepsilon_n, \beta, q, \eta, \frac{L_n}{a}, \frac{a}{R}, \ldots\right) \tag{1.7}$$

where ε_n is the ratio $2L_n/L_B$, η is the ratio L_n/L_T (one for each species), ρ is the gyroradius, L_J is the scale length of variable J (e.g. $L_T = -T/(\partial T/\partial r)$) and α is a parameter that characterizes the transport. Thus, while all the parameters that are arguments of f are dimensionless parameters that do not depend on the size of the system, the factor in front will determine the scaling with size and magnetic field. Two types of transport in particular have been discussed in the literature. They are: Bohm diffusion corresponding to $\alpha = 0$, and gyro-Bohm diffusion corresponding to $\alpha = 1$. In Bohm diffusion the transport is due to global modes that depend on the system size, while in gyro-Bohm diffusion local modes that primarily depend on the gyroradius are responsible for the transport. Clearly, gyro-Bohm diffusion gives a more optimistic extrapolation to larger systems with stronger magnetic fields.

A recently obtained scaling of the confinement time in the H mode is [1.14]

$$\tau_{E\mathrm{aux}} = 0.053 \; I^{1.06} B^{0.32} \kappa^{0.66} A^{0.41} n_e^{0.17} a^{-0.11} R^{1.68} P_{\mathrm{tot}}^{-0.67} \tag{1.8}$$

where κ is the ellipticity (elongation) of the cross-section, A is the mass number and P_{tot} is the total heating power (the remaining variables have been defined above).

This power dependence can be attributed to the unfavourable temperature scaling of the transport caused by the trapped electron and η_i modes, in combination with transport coefficients that grow with temperature gradient. In this new operating regime for beam-heated tokamaks [1.11] the confinement time increased to nearly the ohmic value (roughly a factor of 2). It occurred for a comparatively high electron edge temperature (about 600 eV a few centimetres from the edge) and relatively flat density profiles. In other experiments (Alcator C), an improved confinement was observed when the machine was fuelled by frozen deuterium pellets in the centre rather than by gas puffing at the edge. Since this procedure leads to more peaked density profiles, and hence lower η_e and η_i, it supports our previous interpretation of the transport as due to η_i modes and trapped electron or microtearing modes.

1.3 Discussion

Strong progress has recently been made in fusion research both experimentally and theoretically. Experimentally, this has been made through the discovery of new improved confinement regimes and through extension of tokamak operating regimes through plasma shaping. Theory has made progress largely because of the strong development of computers and simulation techniques. A particularly obvious example is the development of nonlinear gyrokinetic simulations [1.15]. The main new feature is the possibility of including the full toroidal effects.

References

[1.1] Lawson J D 1957 *Proc. Phys. Soc.* B **70** 6

[1.2] McNally J R Jr 1977 *Nucl. Fusion* **17** 1273

[1.3] Stacey W M Jr 1981 *Fusion Plasma Analysis* (New York: Wiley)

[1.4] Troyon F, Gruber R, Saurenmann H, Semenzato S and Succi S 1984 *Plasma Phys. Control. Fusion* **26** 209

[1.5] Gaudreau M *et al* 1977 *Phys. Rev. Lett.* **39** 1266

[1.6] Kadomtsev B B and Pogutse O P 1969 *Sov. Phys.–Dokl.* **14** 470

[1.7] Drake J F, Gladd N T, Liu C S and Chang C L 1980 *Phys. Rev. Lett.* **44** 994

[1.8] Rudakov L I and Sagdeev R Z 1961 *Sov. Phys.–Dokl.* **6** 415

[1.9] Manheimer W M and Antonsen T M 1979 *Phys. Fluids* **22** 957

[1.10] Coppi B 1980 *Comment. Plasma Phys. Control. Fusion* **5** 261

[1.11] Wagner F *et al* 1982 *Phys. Rev. Lett.* **53** 1453

[1.12] Kadomtsev B B 1975 *Sov. J. Plasma Phys.* **1** 295

[1.13] Connor J W and Taylor J B 1977 *Nucl. Fusion* **17** 1047

[1.14] Kardaun O *et al* 1992 ITER: analysis of the H-mode confinement and threshold databases *Proc. 14th Int. Conf. on Plasma Physics and Controlled Nuclear Fusion Research* vol 3 (Vienna: IAEA) p 251

[1.15] Parker S E *et al* 1994 *Phys. Plasmas* **1** 1461

Chapter 2

Fluid Equations for Low Frequency Phenomena

In inhomogeneous magnetized plasmas, low frequency modes with frequency ω much less than the cyclotron frequency Ω_c are considered to be the most dangerous ones for the establishment of quasi-stationary high beta plasma states, necessary for the realization of thermonuclear fusion [2.1–2.15]. The main common feature of these modes is that they are characterized by $k_{\parallel} \ll k_{\perp}$. This means that these modes have a very slow variation along the magnetic field and may be denoted quasi-flute modes. As will be demonstrated in the next section, a vorticity is associated with the leading order perpendicular velocity perturbation. We may thus refer to these modes as vortex modes.

2.1 Fluid Motion Perpendicular to the Magnetic Field

The fluid approximation in the plane perpendicular to the magnetic field is assumed to be due to a strong magnetization of the particles. The lowest order kinetic correction is then the ion finite Larmor radius effect, which can be obtained from inertial diamagnetic effects and stress tensor drifts, i.e., from the part of the momentum equation that couples to the energy equation. Since most effects characteristic of inhomogeneous plasmas can be obtained without temperature perturbations and background gradients we shall initially neglect these. For the fluid approximation in the direction along the magnetic field, the conditions are more restrictive and will basically be fulfilled when the phase velocity of the perturbation is zero or much larger than the thermal velocity, as discussed in section 2.4.

The region $\omega \ll \Omega_{ci}$, $\Omega_{ci} = eB/m_i$ is particularly simple to treat from the fluid motion point of view, since here the equation of motion may be solved by an algebraic iterative method. We write the equation of motion as

$$\frac{\partial \boldsymbol{v}_j}{\partial t} + (\boldsymbol{v}_j \cdot \nabla)\boldsymbol{v}_j = \frac{q_j}{m_j}(\boldsymbol{E} + \boldsymbol{v}_j \times \boldsymbol{B}) - \frac{1}{m_j n_j}(\nabla p_j + \nabla \cdot \boldsymbol{\pi}_j) + \boldsymbol{g}_j. \quad (2.1)$$

Here we introduced the isotropic pressure $p_j = n_j T_j$, the anisotropic stress tensor $\boldsymbol{\pi}_j$ and an external force represented by a gravitational acceleration $\boldsymbol{g}_j$. Assuming an electrostatic approximation where $\boldsymbol{B} = B_0\hat{\boldsymbol{z}}$, we obtain by taking the vector product of equation (2.1) with $\hat{\boldsymbol{z}}$

$$\left(\frac{\partial}{\partial t} + \boldsymbol{v}_j \cdot \nabla\right) \hat{\boldsymbol{z}} \times \boldsymbol{v}_j = \frac{q_j}{m_j}\{\hat{\boldsymbol{z}} \times \boldsymbol{E} + B_0[\boldsymbol{v}(\hat{\boldsymbol{z}} \cdot \hat{\boldsymbol{z}}) - \hat{\boldsymbol{z}}(\hat{\boldsymbol{z}} \cdot \boldsymbol{v})]\}$$
$$- \frac{1}{m_j n_j}\hat{\boldsymbol{z}} \times (\nabla p_j + \nabla \cdot \boldsymbol{\pi}_j) + \hat{\boldsymbol{z}} \times \boldsymbol{g}_j.$$

By writing $\mathrm{d}/\mathrm{d}t = \partial/\partial t + \boldsymbol{v}_j \cdot \nabla$ we now find

$$\boldsymbol{v}_\perp = \frac{1}{B_0}(\boldsymbol{E} \times \hat{\boldsymbol{z}}) + \frac{1}{B_0}\frac{m_j}{q_j}\frac{\mathrm{d}}{\mathrm{d}t}(\hat{\boldsymbol{z}} \times \boldsymbol{v}) + \frac{1}{q_j n_j B_0}\hat{\boldsymbol{z}} \times (\nabla p_j + \nabla \cdot \boldsymbol{\pi}_j) - \frac{m_j}{q_j B_0}\hat{\boldsymbol{z}} \times \boldsymbol{g}_j.$$

Here the first term is the $E \times B$ drift. The second term is the polarization drift and contains the perturbed velocity. If we assume the $E \times B$ drift to be the dominating part of the perturbed velocity we may substitute it into the polarization drift. We then write the perpendicular velocity as

$$\boldsymbol{v}_\perp = \boldsymbol{v}_E + \boldsymbol{v}_p + \boldsymbol{v}_* + \boldsymbol{v}_\pi + \boldsymbol{v}_g \tag{2.2}$$

where

$$\boldsymbol{v}_E = \frac{1}{B}(\boldsymbol{E} \times \hat{\boldsymbol{z}}) \tag{2.3}$$

$$\boldsymbol{v}_{pj} = \frac{1}{B_0\Omega_{cj}}\left[\frac{\partial}{\partial t}\boldsymbol{E} + (\boldsymbol{v} \cdot \nabla)\boldsymbol{E}\right] \tag{2.4}$$

$$\boldsymbol{v}_{*j} = \frac{\hat{\boldsymbol{z}} \times \nabla p_j}{q_j n_j B_0} \tag{2.5}$$

$$\boldsymbol{v}_{\pi j} = \frac{\hat{\boldsymbol{z}} \times \nabla \cdot \boldsymbol{\pi}_j}{q_j n_j B_0} \tag{2.6}$$

and

$$\boldsymbol{v}_{gj} = -(\hat{\boldsymbol{z}} \times \boldsymbol{g})/\Omega_{cj}. \tag{2.7}$$

Here $\Omega_{cj} = q_j B_0/m_j$ and $\nabla_\perp = \hat{\boldsymbol{x}}\partial/\partial x + \hat{\boldsymbol{y}}\partial/\partial y$.

We note that the assumption $v_{p_j} \ll v_E$ is consistent with the assumption $\omega \ll \Omega_{cj}$. For this approximation to be generally valid we must have $\omega \ll \Omega_{ci}$, where Ω_{ci} is the cyclotron frequency of the ions. We note here that $(\nabla \times \boldsymbol{v}_E)_\parallel = (c/B_0)\Delta_\perp\phi\hat{\boldsymbol{z}}$, i.e., we have a vorticity associated with the $E \times B$ drift. This drift is the same for electrons and ions. The velocity $\boldsymbol{v}_{*j}$ is the diamagnetic drift velocity. This is a pure fluid velocity and is not a particle drift. It fulfils (as follows from appendix 2, for homogeneous B and $\hat{z}$)

$$\nabla \cdot (n\boldsymbol{v}_{*j}) = 0. \tag{2.8}$$

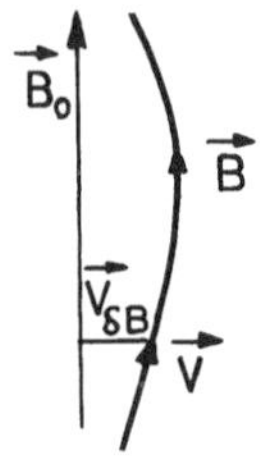

Figure 2.1. Drift due to field line bending.

Analogously

$$\nabla \cdot \boldsymbol{v}_E = 0. \tag{2.9}$$

The velocity $\boldsymbol{v}_\pi$ is due to the stress tensor $\boldsymbol{\pi}$, which contains a viscosity part $\boldsymbol{\pi}_v$ and a finite Larmor radius part $\boldsymbol{\pi}_l$. The finite Larmor radius part $\boldsymbol{v}_{\pi l}$ of $\boldsymbol{v}_\pi$ fulfils

$$\nabla \cdot (n\boldsymbol{v}_{\pi l}) \sim k_\perp^2 \rho^2 n_0 |\nabla \ln n_0| v_\perp.$$

This contribution may be neglected if $k_\perp^2 \rho^2 \ll 1$.

2.2 Electromagnetic Drifts

When the magnetic field is perturbed, i.e.

$$\boldsymbol{B} = B_0 \hat{\boldsymbol{z}} + \delta \boldsymbol{B}$$

we have

$$\hat{\boldsymbol{z}} \times (\boldsymbol{v} \times \boldsymbol{B}) = B_\parallel \boldsymbol{v} - v_\parallel \boldsymbol{B} = B_\parallel \boldsymbol{v}_\perp - v_\parallel \delta \boldsymbol{B}_\perp$$

where $\parallel$ and $\perp$ refer to $\hat{\boldsymbol{z}}$ and $B_\parallel = B_0 + \delta B_\parallel$. We now solve for $\boldsymbol{v}_\perp$ in the same way as in the preceding section, dividing by $B_\parallel$. This gives us back the same drifts as before, but now with B_0 replaced by $B_\parallel$ in the denominator and also a new drift of the form

$$\boldsymbol{v}_{\delta B} = v_\parallel \frac{\delta \boldsymbol{B}_\perp}{B_\parallel}. \tag{2.10}$$

This drift is caused by projection of the parallel velocity in the perpendicular direction due to field line bending (see figure 2.1). In the fluid description this drift is nonlinear unless we have a background drift along $\hat{\boldsymbol{z}}$. Introducing the potentials ϕ and $\boldsymbol{A}$ by $\boldsymbol{E} = -\nabla\phi - \partial \boldsymbol{A}/\partial t$, $\delta \boldsymbol{B} = \nabla \times \boldsymbol{A}$ we have

$$\boldsymbol{v}_E = \frac{1}{B}(\hat{\boldsymbol{z}} \times \nabla\phi) + \frac{1}{B}\left(\hat{\boldsymbol{z}} \times \frac{\partial \boldsymbol{A}}{\partial t}\right) \tag{2.11}$$

and we note that in a system without curvature (B and $\hat{z}$ homogeneous) we have

$$\nabla \cdot \boldsymbol{v}_E = -\frac{1}{B_\parallel}\frac{\partial}{\partial t}\delta B_\parallel. \tag{2.12}$$

Since a divergence of the velocity is equivalent to compressibility ($\mathrm{d}n/\mathrm{d}t = \partial n/\partial t + \boldsymbol{v} \cdot \nabla n = -n\nabla \cdot \boldsymbol{v}$), we conclude that for processes where $\boldsymbol{v}_E$ is the dominant velocity, compressibility is associated with a perturbation in $B_\parallel$. Since for low frequency processes $\delta B_\parallel / B \sim \beta \delta P / P$, where $\beta = 2\mu_0 P / B_0^2$ (see equation (5.15)), the divergence of $\boldsymbol{v}_E$ may for such processes also be considered as a finite β effect. Since β typically is small ($\leq 5\%$ for present day fusion experiments), a very common approximation is to neglect $\nabla \cdot \boldsymbol{v}_E$ in the absence of curvature.

Since for $\omega \ll \Omega_c$, $\boldsymbol{v}_E$ is the dominant particle drift ($\boldsymbol{v}_*$ is a pure fluid drift, as will be discussed in the next section), the perpendicular ion and electron motion will be the same to lowest order so that the whole plasma obtains the same displacement. Since the induction electric field along the magnetic field has a tendency to counteract perturbations, it tends to reduce the electric field along $\boldsymbol{B}$. In the ideal MHD limit this gives $\boldsymbol{E} \cdot \boldsymbol{B} = 0$. Then the magnetic field is displaced in the same way as the plasma so that the magnetic flux through a given surface is conserved.

2.3 Interpretation of Drifts

The drifts derived here are fluid drifts. They may differ from actual particle or guiding centre drifts and these differences are sometimes sources of confusion. The reason for the differences is that the fluid picture averages particle velocities at a point, regardless of where the guiding centres are located, while the particle drifts are obtained by first averaging over the gyromotion, thus identifying a particle with its guiding centre. The diamagnetic drift is due to a pressure gradient (figure 2.2).

Since more particles or particles with a larger velocity are moving upwards we obtain a net average velocity, although the guiding centres do not move. The diamagnetic drift is thus a fluid drift but not a particle drift. This drift is in opposite directions for electrons and ions and produces the diamagnetic current, which in turn balances the plasma pressure through the $\boldsymbol{j} \times \boldsymbol{B}$ force (compare section 5.2.1). It is also the source of the magnetic field variation across a plasma boundary as indicated by its name. The diamagnetic drift does not cause charge separation as shown by equation (2.8). This case is in contrast to the ∇B and curvature drifts, which are particle drifts but not fluid drifts. In the presence of a perpendicular magnetic field gradient the Larmor radius is larger on one side than on the other, leading to influence from more particles from that side on the average velocity. The effect compensates exactly the ∇B drift (figure 2.3).

A similar argument holds for the curvature drift. As it turns out, the difference between the fluid current and the particle current is the magnetization

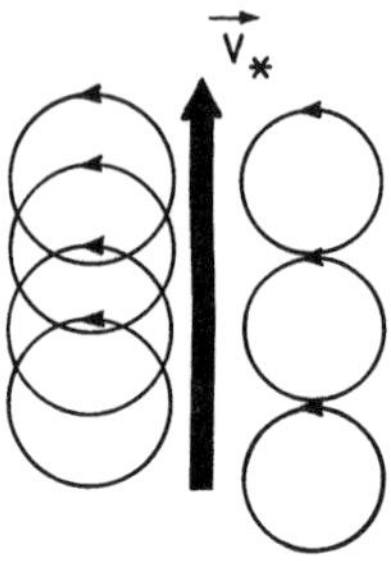

Figure 2.2. The diamagnetic drift.

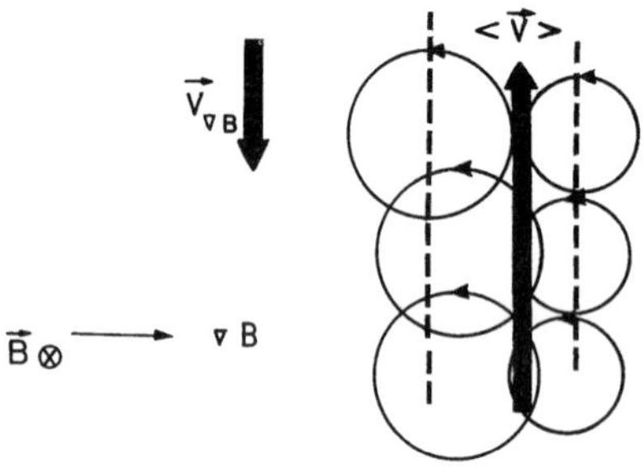

Figure 2.3. Fluid drift in an inhomogeneous magnetic field.

current

$$j_m = \nabla \times M \tag{2.13}$$

where

$$M = \frac{q}{|q|}(n_e\mu_e + n_i\mu_i)e_{\parallel} \tag{2.14}$$

and where

$$\mu_j = \frac{m_j v_{\perp j}^2}{2B_0} \rightarrow \frac{T_j}{B}$$

is the magnetic moment, and $e_{\parallel} = B/B$.

An important property of j_m is that

$$\nabla \cdot j_m = 0. \tag{2.15}$$

This property is very useful since eigenvalue equations and dispersion relations are frequently obtained from the low frequency relation

$$\nabla \cdot j = 0.$$

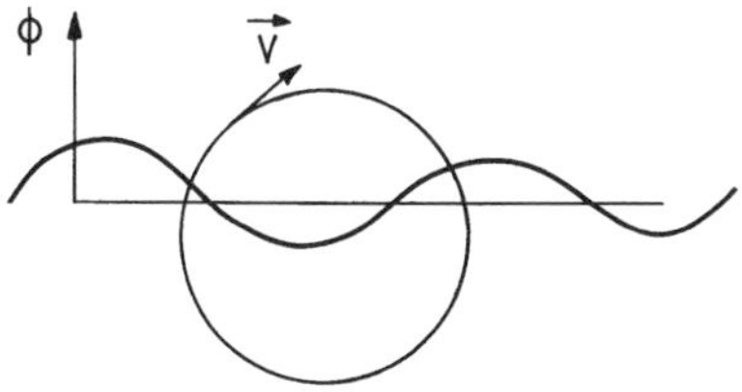

Figure 2.4. Finite gyro orbit.

We are thus free to substitute either fluid only or particle only drifts into this equation with the same result. This is particularly useful when finite Larmor radius effects are included, in which case particle drifts are easier to obtain than fluid drifts.

The remaining drifts (except $\boldsymbol{v}_\pi$) are the same for the fluid and particle descriptions. We may note here that since $\boldsymbol{v}_E$ is the same for electrons and ions it does not give rise to a current. The polarization drift on the other hand may be regarded as a correction to the $E \times B$ drift when the E-field is time dependent. Due to the strong difference in inertia between electrons and ions, the electron polarization drift is usually neglected. We note also a strong similarity between the polarization drift caused by a time variation of $\boldsymbol{E}$ and the finite Larmor radius drift, which is due to a space variation of $\boldsymbol{E}$ (figure 2.4). This is because a gyrating particle has no way of deciding if the variation in $\boldsymbol{E}$ which it experiences along its orbit originates in a time or space variation of $\boldsymbol{E}$.

2.4 Conservation of Charge Density and Quasi-Neutrality

We shall henceforth make frequent use of quasi-neutrality and conservation of charge density. For low frequencies these concepts are related by the neglect of the displacement current.

The Poisson equation is written

$$\Delta\phi = \frac{e}{\varepsilon_0}(n_e - n_i).$$

Writing $\Delta\phi = -k^2\phi$ we may rewrite it as

$$\frac{n_i - n_e}{n_0} = \frac{\varepsilon_0 T_e}{n_0 e^2} k^2 \frac{e\phi}{T_e} = \frac{k^2}{k_D^2}\frac{e\phi}{T_e}$$

where n_0 is the unperturbed background density, and we have introduced $k_D^2 = n_0 e^2 / T_e \varepsilon_0$. For perturbations fulfilling $e\phi/T_e \leq 1$, we find that the relative density difference is small if $k^2 \leq k_D^2$. Usually, as will be shown later, the perturbation in density fulfils $\delta n/n_0 \sim e\phi/T_e$ for low frequency modes.

Thus, replacing n_0 by $\delta n(T_e/e\phi)$ we find the estimate

$$\frac{\delta n_i - \delta n_e}{\delta n} \sim \frac{k^2}{k_D^2}. \tag{2.16}$$

Here δn represents a measure of the average perturbation in density $(\delta n_i + \delta n_e)/2$. We then conclude that for $k^2 \ll k_D^2$ also the relative difference in the perturbation of ion and electron densities is small. To first order of approximation we then replace the Poisson equation by $\delta n_i \approx \delta n_e$. Multiplying the electron and ion continuity equations by the particle charges and adding them together we obtain the equation for conservation of charge density

$$\frac{\partial \rho}{\partial t} + \nabla \cdot j = 0.$$

With $\rho = e(n_i - n_e) \sim 0$, if we assume quasi-neutrality, we obtain the condition

$$\nabla \cdot j = 0. \tag{2.17}$$

This condition is really just a low frequency condition since $\rho \sim 0$ is a condition the plasma can maintain at low frequencies. The additional time derivative in the continuity equation just enhances this fact.

The condition (2.17) can also be obtained by taking the divergence of Ampère's law, neglecting the displacement current. In the following we shall usually derive dispersion or eigenvalue equations by substituting the particle drifts into condition (2.17), assuming quasi-neutrality. In a magnetized plasma, the particle dynamics are usually very different in directions parallel to and perpendicular to the magnetic field. In the perpendicular direction particles are localized by the magnetic field and a fluid description is usually valid. In the parallel direction, however, the conditions for a fluid model are much more difficult to fulfil. There are essentially three situations where we may use a fluid description in this direction. They are

(a) $\omega \gg k_\parallel v_{\text{th}}$;
(b) $k_\parallel \ll 1/l_f$, $\nu \gg \omega$;
(c) $\omega/k_\parallel \sim 0$ and symmetric distribution.

Here, (a) corresponds to the cold plasma approximation, i.e., for fast enough processes the plasma particles appear to be stationary and thus localized. In (b), when the mean free path l_f is much shorter than a wavelength the particles are not free to maintain communication between different parts of our perturbation and are thus localized. In (c), for zero phase velocity and symmetric distribution function the nonlocal effects cancel. This is the situation, for example, for the electrons in ion acoustic waves, and as we shall see in the following, also in drift waves. The fluid model here, however, holds only to zero order in $\omega/k_\parallel v_t$.

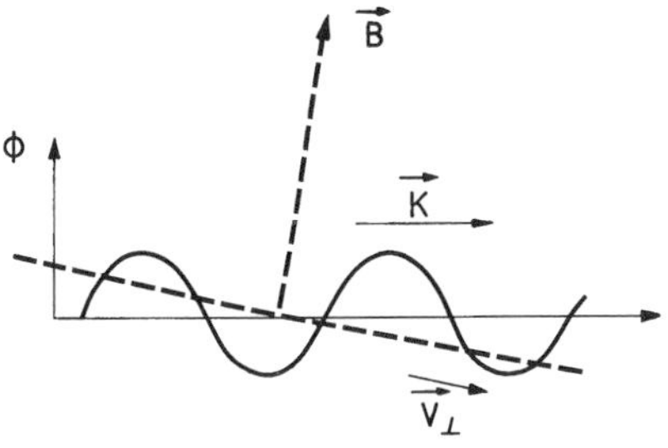

Figure 2.5. Propagation at near right angles to the magnetic field.

Obviously, the above conditions will, in general, apply differently for electrons and ions. A typical situation in the following is when (a) holds for ions and (c) holds for electrons.

It is now natural to divide the condition (2.17) into parallel and perpendicular parts. Instabilities are typically generated by differences in electron and ion perpendicular drifts, while electron motion along the field lines has a stabilizing influence by short-circuiting the charge separation (figure 2.5).

Effects that impede the free electron motion along the field lines are then destabilizing. Such effects may be electron–ion collisions, electron Landau damping and magnetic induction. We shall see examples of all three of these below.

2.5 Simple Examples

As simple applications of the low frequency fluid expansion we shall now consider two wave types in homogeneous plasmas: the shear Alfvén wave and the magnetosonic wave. Substituting the representation $\boldsymbol{B} = \nabla \times \boldsymbol{A}$ into the Ampère law, neglecting displacement current, and using the gauge condition $\nabla \cdot \boldsymbol{A} = 0$ we may write the parallel component as

$$j_{\parallel} = -\frac{1}{\mu_0} \Delta A_{\parallel}. \tag{2.18}$$

Assuming perfect conductivity along the field lines we have the condition $E_{\parallel} = 0$ leading to the relation

$$A_{\parallel} = \frac{k_{\parallel}}{\omega} \phi. \tag{2.19}$$

Taking the divergence of the Ampère law we now have equation (2.17)

$$\nabla \cdot j = 0.$$

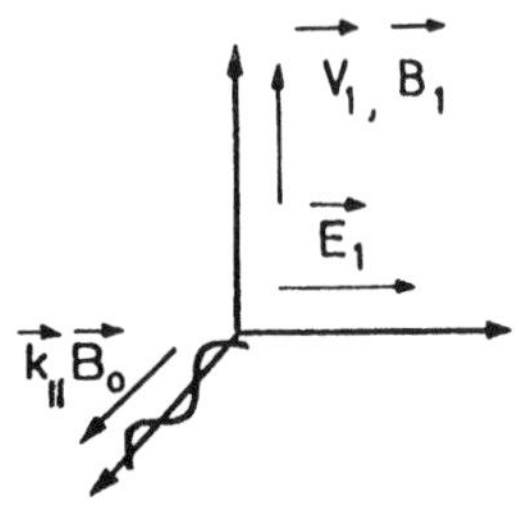

Figure 2.6. Shear Alfvén wave.

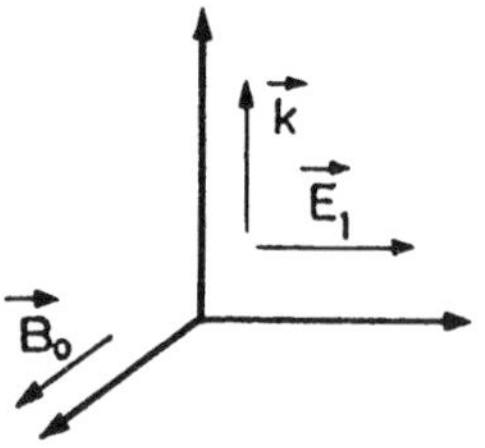

Figure 2.7. Magnetosonic wave.

Then substituting equation (2.18) for $j_\parallel$ in equation (2.17) and using the ion polarization drift in $j_\perp$ we obtain (the $E \times B$ drifts cancel and the diamagnetic drifts do not contribute):

$$-\frac{1}{\mu_0} e_\parallel \cdot \nabla \Delta \frac{k_\parallel}{\omega} \phi = en\nabla \cdot \left(\frac{1}{B_0 \Omega_c} \frac{\partial}{\partial t} \nabla_\perp \phi \right).$$

Assuming that $k_\perp \gg k_\parallel$ and $\delta B_\parallel = 0$ we obtain the dispersion relation

$$\omega^2 = k_\parallel^2 v_A^2 \tag{2.20}$$

where $v_A = B_0/\sqrt{\mu_0 n_0 m_i}$ is the Alfvén velocity. The reason for including only $A_\parallel$ is that the shear Alfvén wave only bends the field line (figure 2.6). As it turns out, however, we can obtain equation (2.20) for a general ratio of $k_\parallel/k_\perp$ by including the effects of $\boldsymbol{A}_\perp$.

The magnetosonic or compressional Alfvén wave typically has a considerably higher frequency than the shear Alfvén wave. This mode is associated with parallel perturbations in $\boldsymbol{B}$, which requires higher energy than the line bending. It propagates perpendicular to $\boldsymbol{B}_0$ and compresses the plasma in the direction of propagation by means of the $E \times B$ drift. When the magnetic field tends to zero it turns into a regular ion acoustic wave (figure 2.7).

For the magnetosonic branch we use the perpendicular correspondence of equation (2.18)

$$j_\perp = -\frac{1}{\mu_0}\Delta \boldsymbol{A}_\perp .$$

In the limit $T_e \gg T_i$, corresponding to regular ion acoustic waves, we have

$$j_\perp \approx en\boldsymbol{v}_{pi} - en\boldsymbol{v}_{*e}. \tag{2.21}$$

Thus neglecting temperature perturbations and linearizing we obtain

$$-\frac{1}{\mu_0}\Delta \boldsymbol{A}_\perp = \frac{en_0}{B_0\Omega_{ci}}\frac{\partial \boldsymbol{E}_\perp}{\partial t} + T_e\frac{1}{B}(\boldsymbol{e}_\parallel \times \nabla \delta n_e). \tag{2.22}$$

Neglecting polarization drifts and parallel motion ($k_\parallel = 0$), we obtain from the continuity equation

$$\frac{\delta n_e}{n} = \frac{\delta B_\parallel}{B_0}. \tag{2.23}$$

Substituting equation (2.23) into equation (2.22) and using appendix 2 we obtain with $\nabla \cdot \boldsymbol{A} = 0$ and $\hat{\boldsymbol{z}} \cdot \nabla = 0$

$$\frac{\partial \boldsymbol{E}_\perp}{\partial t} = -\Delta \boldsymbol{A}_\perp (v_A^2 + c_s^2)$$

where $c_s^2 = T_e/m_i$. Now, since $\boldsymbol{k} \cdot \boldsymbol{E} = 0$ the electrostatic part of $\boldsymbol{E}$ vanishes and we immediately obtain the dispersion relation

$$\omega^2 = k_\perp^2 (v_A^2 + c_s^2). \tag{2.24}$$

Here the second part is a finite β effect ($c_s^2/v_A^2 \approx \frac{1}{2}\beta_e$) and is in accordance with equation (2.23) associated with compression of the plasma.

The Alfvén velocity v_A is one of the fundamental parameters for low frequency phenomena and also enters the low frequency perpendicular dielectric constant.

2.6 The Energy Equation

The highest order moment equation that we shall make use of is the energy equation. It is most commonly written as an equation for the pressure variation

$$\frac{3}{2}\left(\frac{\partial}{\partial t} + \boldsymbol{v}_j \cdot \nabla\right) P_j + \frac{5}{2} P_j \nabla \cdot \boldsymbol{v}_j = -\nabla \cdot \boldsymbol{q}_j + \sum_{i=j} Q_{ji} \tag{2.25}$$

where $\boldsymbol{q}_j$ is the heat flux and Q_{ij} is the heat transferred from species i to species j by means of collisions. This energy exchange typically contains effects like ohmic heating and temperature equilibrium terms. It will be neglected in the following. The heat flux $\boldsymbol{q}$ is for the collision dominated case ($\lambda \gg l_f$) according to Braginskii [2.4]

$$\boldsymbol{q}_j = 0.71 n_j T_j \boldsymbol{U}_\parallel - \kappa_{\parallel j}\nabla_\parallel T_j - \kappa_{\perp j}\nabla_\perp T + \boldsymbol{q}_{*j} + \frac{3}{2}\nu_j \frac{n_j T_j}{\Omega_{cj}}(\boldsymbol{e}_\parallel \times \boldsymbol{U}) \qquad (2.26)$$

where $\boldsymbol{U}$ is the relative velocity between species j and i. The thermal conductivities for electrons are given by

$$\kappa_{\parallel e} = 3.16\frac{n_e T_e}{m_e \nu_e} \qquad \text{and} \qquad \kappa_{\perp e} = 4.66\frac{n_e T_e \nu_e}{m_e \Omega_{ce}^2}$$

and for ions by

$$\kappa_{\parallel i} = 3.9\frac{n_i T_i}{m_i \nu_i} \qquad \text{and} \qquad \kappa_{\perp i} = 2\frac{n_i T_i \nu_i}{m_i \Omega_{ci}^2}$$

and

$$\boldsymbol{q}_{*j} = \frac{5}{2}\frac{n_j T_j}{m_j \Omega_{cj}}(\boldsymbol{e}_\parallel \times \nabla T_j). \qquad (2.27)$$

The unit vector $\boldsymbol{e}_\parallel$ is here defined by $\boldsymbol{e}_\parallel = \boldsymbol{B}/B$, i.e., along the total magnetic field.

If we neglect the full right-hand side of equation (2.26) we obtain the adiabatic equation of state for three dimensional motion, i.e.

$$\frac{\mathrm{d}}{\mathrm{d}t}(Pn^{5/3}) = 0 \qquad (2.28)$$

which holds for processes that are so rapid that the heat flux does not have time to develop. When $\nabla \cdot \boldsymbol{v} = 0$, which is a rather common situation, the pressure perturbation can be taken as being due to convection in a background gradient. This will be further discussed later.

Another usual form of the energy equation is that obtained after subtracting the continuity equation. It may be written as

$$\frac{3}{2}n_j\left(\frac{\partial}{\partial t} + \boldsymbol{v}_j \cdot \nabla\right)T_j + P_j \nabla \cdot \boldsymbol{v} = -\nabla \cdot \boldsymbol{q}_j. \qquad (2.29)$$

Equations (2.25) and (2.29) are fluid equations and the velocities thus contain the diamagnetic drifts. As it turns out, these drifts cancel in a way similar to that in the momentum equation, but now due to the heat flow terms, i.e.

$$\tfrac{3}{2}n\boldsymbol{v}_* \cdot \nabla T - T\boldsymbol{v}_* \cdot \nabla n = \tfrac{5}{2}n\boldsymbol{v}_* \cdot \nabla T$$

$$\tfrac{3}{2}n_j\boldsymbol{v}_{*j} \cdot \nabla T_j - T_j\boldsymbol{v}_{*j} \cdot \nabla n_j = -\nabla \cdot \boldsymbol{q}_{*j}. \qquad (2.30)$$

We can then write the energy equation in the form

$$\frac{3}{2}n_j\left(\frac{\partial}{\partial t} + \boldsymbol{v}_{gcj} \cdot \nabla\right)T_j - T_j\left(\frac{\partial n_j}{\partial t} + \boldsymbol{v}_{gcj} \cdot \nabla n_j\right) = -\nabla \cdot \boldsymbol{q}_{gcj} \qquad (2.31)$$

where $\boldsymbol{v}_{gc}$ is defined here as the guiding centre part of the fluid velocity (i.e., without curvature and ∇B drifts), and $\boldsymbol{q}_{gcj}$ is $\boldsymbol{q}_j$ defined in equation (2.26) but without the diamagnetic part $\boldsymbol{q}_{*j}$.

Equation (2.31) shows that the relevant convective velocity in the energy equation is the guiding centre part of the fluid velocity. The term coming from $\nabla \cdot \boldsymbol{v}_j$ is

$$\frac{\partial n_j}{\partial t} + \boldsymbol{v}_{gcj} \cdot \nabla n_j = -n\nabla \cdot \boldsymbol{v}_{gcj} - \nabla \cdot (n\boldsymbol{v}_{*j})$$

where $\nabla \cdot (n\boldsymbol{v}_{*j})$ is a pure curvature effect. From this it also follows that the convective velocity in equation (2.28) does not contain the diamagnetic drift.

Another useful equation of state may be obtained at low frequencies and small collision rates for electrons. In this case, the energy equation is dominated by the the $\nabla \cdot \boldsymbol{q}$ term so that the lowest order equation of state is $\boldsymbol{q} = 0$ or $\kappa_\parallel \nabla_\parallel T = 0$. Now $\nabla_\parallel = (1/B)(\boldsymbol{B}_0 + \delta\boldsymbol{B}) \cdot \nabla$, so that after linearization

$$\boldsymbol{B}_0 \cdot \nabla T_{1j} + \delta\boldsymbol{B} \cdot \nabla T_{0j} = 0. \tag{2.32}$$

If the perpendicular perturbation in $\boldsymbol{B}$ is represented by a parallel vector potential we obtain the equation of state

$$T_{1j} = -\eta_j \frac{\omega_{*j}}{k_\parallel} q_j A_\parallel \tag{2.33}$$

where $\eta_j = \mathrm{d}\ln T_j/\mathrm{d}\ln n_j$.

Although the above expression for $\boldsymbol{q}$ has been derived by assuming domination of collisions along $\boldsymbol{B}$ ($\lambda \gg l_f$), the equation of state (2.33) can also be used to reproduce the electron density response in the limit $\omega \ll k_\parallel v_\parallel$ obtained from the Vlasov equation. The reason for this is that it arises as a limiting case that does not depend on the explicit form of $\kappa_\parallel$.

With regard to the cancellation of the diamagnetic drifts, this effect is very important for vortex modes since the perturbed part of $\boldsymbol{v}_*$ is typically of the same order as $\boldsymbol{v}_E$. The application of equation (2.28) for such modes thus depends strongly on this cancellation and the relevant convective velocity in $\mathrm{d}/\mathrm{d}t$ is the guiding centre part of the fluid velocity.

2.7 Finite Larmor Radius Effects in a Fluid Description

Up to now we have neglected diamagnetic contributions to the polarization drift and the stress tensor drift. As it turns out these are related to finite Larmor radius (FLR) effects. We shall show here how the lowest order FLR effects can be obtained by a systematic inclusion of these terms.

We shall, for simplicity, initially neglect temperature gradients and temperature perturbations. This leads to the relation

$$\nabla \cdot \boldsymbol{v}_* = \frac{T}{qB} \nabla \cdot (\hat{\boldsymbol{z}} \times \nabla n/n) = 0. \tag{2.34}$$

Since $\nabla \cdot \boldsymbol{v}_E = 0$ we can use the incompressibility condition $\nabla \cdot \boldsymbol{v} = 0$ to leading order when substituting drifts into $\boldsymbol{v}_p$ and $\boldsymbol{v}_\pi$. We shall also assume large mode numbers, i.e., $k \gg \kappa = |\nabla \ln n_0|$ and $\nabla\kappa = 0$.

From the stress tensor as given by Braginskii we can obtain effects of viscosity related to friction between particles and collisionless gyroviscosity, which is a pure FLR effect. The relevant gyroviscous components are

$$\begin{aligned}\pi_{xy} = \pi_{yx} &= \frac{nT}{2\Omega_c}\left(\frac{\partial v_x}{\partial x} - \frac{\partial v_y}{\partial y}\right) + \frac{1}{4\Omega_c}\left(\frac{\partial q_x}{\partial x} - \frac{\partial q_y}{\partial y}\right)\\ \pi_{yy} = -\pi_{xx} &= \frac{nT}{2\Omega_c}\left(\frac{\partial v_y}{\partial x} + \frac{\partial v_x}{\partial y}\right) + \frac{1}{4\Omega_c}\left(\frac{\partial q_x}{\partial y} + \frac{\partial q_y}{\partial x}\right).\end{aligned} \tag{2.35}$$

Here $\boldsymbol{q}$ is determined by the fluid truncation and will include higher order FLR effects. We note, however, that the part of $\boldsymbol{q}_*$ corresponding to a flux of perpendicular energy is [2.12]:

$$\boldsymbol{q}_*^{\perp} = 2\frac{P_\perp}{m\Omega_c}(\hat{\boldsymbol{z}} \times \nabla T_\perp).$$

We start by including only the density background gradient

$$\begin{aligned}(\nabla \cdot \boldsymbol{\pi})_x &= \frac{\partial \pi_{xx}}{\partial x} + \frac{\partial \pi_{xy}}{\partial y} = -\frac{nT}{2\Omega_c}\Delta v_y - \frac{T}{2\Omega_c}\left(\frac{\partial v_y}{\partial x} + \frac{\partial v_x}{\partial y}\right)\frac{\mathrm{d}n}{\mathrm{d}x}\\ (\nabla \cdot \boldsymbol{\pi})_y &= \frac{\partial \pi_{yx}}{\partial x} + \frac{\partial \pi_{yy}}{\partial y} = \frac{nT}{2\Omega_c}\Delta v_x + \frac{T}{2\Omega_c}\left(\frac{\partial v_x}{\partial x} - \frac{\partial v_y}{\partial y}\right)\frac{\mathrm{d}n}{\mathrm{d}x}.\end{aligned}$$

These equations may be written in a more compact form as

$$\nabla \cdot \boldsymbol{\pi} = \frac{nT}{2\Omega_c}[\hat{\boldsymbol{z}} \times \Delta_\perp \boldsymbol{v} + \kappa(\nabla v_y - \hat{\boldsymbol{z}} \times \nabla v_x)]. \tag{2.36}$$

We now obtain

$$\boldsymbol{v}_\pi = \frac{1}{enB}\hat{\boldsymbol{z}} \times \nabla \cdot \boldsymbol{\pi} = -\frac{1}{4}\rho^2\Delta_\perp \boldsymbol{v} + \frac{1}{4}\rho^2\kappa(\hat{\boldsymbol{z}} \times \nabla v_y + \nabla v_x). \tag{2.37}$$

Here ρ is the gyroradius of a general species. Since we are usually interested in substituting our drifts into the equation $\nabla \cdot \boldsymbol{j} = 0$ we need to calculate expressions of the form $\nabla \cdot (n\boldsymbol{v})$. We then find, including only linear terms in κ

$$\begin{aligned}\nabla \cdot (n\boldsymbol{v}_\pi) &= \boldsymbol{v}_\pi \cdot \nabla n_0 + n_0 \nabla \cdot \boldsymbol{v}_\pi\\ &= -\tfrac{1}{4}\rho^2 \nabla n_0 \cdot \Delta_\perp \boldsymbol{v} - \tfrac{1}{4}\rho^2 n_0 \Delta_\perp \nabla \cdot \boldsymbol{v} + \tfrac{1}{4}\rho^2\kappa n_0 \Delta_\perp v_x.\end{aligned}$$

Here we may use $\nabla \cdot \boldsymbol{v} \approx 0$ to obtain

$$\nabla \cdot (n\boldsymbol{v}_\pi) = -\tfrac{1}{2}\rho^2 \nabla n_0 \cdot \Delta \boldsymbol{v}. \tag{2.38}$$

The polarization drift can be written in the form

$$\boldsymbol{v}_p = \frac{1}{\Omega_c}\left(\frac{\partial}{\partial t} + \boldsymbol{v}\cdot\nabla\right)(\hat{\boldsymbol{z}}\times\boldsymbol{v}).$$

We start by observing that due to our large mode number approximation only perturbed drifts will enter the last $\boldsymbol{v}$. Then, in the linear approximation the $\boldsymbol{v}$ term in the convective derivative can only be a background $\boldsymbol{v}$. The only background $\boldsymbol{v}$ that we are interested in here is the diamagnetic drift. We shall then start by considering the contribution from this term to $\nabla\cdot(n\boldsymbol{v}_p)$. It is

$$\begin{aligned}\frac{n}{\Omega_c}(\boldsymbol{v}_*\cdot\nabla)\nabla\cdot(\hat{\boldsymbol{z}}\times\boldsymbol{v}) &= \frac{n}{\Omega_c}\boldsymbol{v}_*\cdot\nabla\left(\frac{\partial v_x}{\partial y} - \frac{\partial v_y}{\partial x}\right)\\ &= -\frac{1}{2}\kappa n\rho^2\frac{\partial}{\partial y}\left(\frac{\partial v_x}{\partial y} - \frac{\partial v_y}{\partial x}\right).\end{aligned} \tag{2.39}$$

Now adding equations (2.38) and (2.39) we find (with $\nabla n_0 = -\kappa n_0\hat{\boldsymbol{x}}$):

$$\begin{aligned}\nabla\cdot(n\boldsymbol{v}_\pi) + \nabla\cdot\left[\frac{n}{\Omega_c}(\boldsymbol{v}_*\cdot\nabla)(\hat{\boldsymbol{z}}\times\boldsymbol{v})\right] &= \frac{1}{2}\rho^2\kappa n\Delta v_x\\ -\frac{1}{2}\rho^2\kappa n\left(\frac{\partial^2 v_x}{\partial y^2} - \frac{\partial^2 v_y}{\partial y\partial x}\right) &= \frac{1}{2}\rho^2\kappa n\frac{\partial}{\partial x}\nabla\cdot\boldsymbol{v} = 0.\end{aligned} \tag{2.40}$$

We thus find that *convective diamagnetic contributions to $\nabla\cdot(n\boldsymbol{v}_p)$ are exactly cancelled by the stress tensor contribution $\nabla\cdot(n\boldsymbol{v}_\pi)$.*

This result can easily be understood from a physical point of view since the diamagnetic drift is not a particle drift and cannot transfer information by convection. *We now have the general result*

$$\nabla\cdot[n(\boldsymbol{v}_p + \boldsymbol{v}_{pi})] = \nabla\cdot\left[\frac{n}{\Omega_c}\frac{\partial}{\partial t}(\hat{\boldsymbol{z}}\times\boldsymbol{v})\right]. \tag{2.41}$$

It is also interesting to compare equation (2.41) with equation (2.8). In order to obtain a result corresponding to equation (2.8) for the gyroviscous part of the stress tensor drift $\boldsymbol{v}_\pi$ it is necessary to add the convective diamagnetic parts of the polarization drift, which are of the same order in $k^2\rho^2$. We may thus consider equation (2.41) to express the same kind of physics as equation (2.8), but for drifts that are first order in the FLR parameter $k^2\rho^2$. Since equation (2.8) is no longer true in the presence of curvature (compare equation (4.20)), the same is expected for equation (2.41).

The leading order linear contributions are

$$\frac{n}{\Omega_{ci}}\frac{\partial}{\partial t}\nabla\cdot(\hat{\boldsymbol{z}}\times\boldsymbol{v}_E) = -\frac{1}{2}n\rho_i^2\frac{\partial}{\partial t}\Delta\frac{e\phi}{T_i}$$

and

$$\frac{n}{\Omega_{ci}}\frac{\partial}{\partial t}\nabla\cdot(\hat{\boldsymbol{z}}\times\boldsymbol{v}_{*i}) = -\frac{1}{2}\rho_i^2\frac{\partial}{\partial t}\Delta\delta n$$

where only the perturbation in density contributes.

We now have to specialize further to a particular density response. For flute modes, which are of particular interest in this context, the simplest leading order density perturbation is the convective, i.e.

$$\frac{\delta n}{n} = \frac{\omega_{*e}}{\omega}\frac{e\phi}{T_e}. \tag{2.42}$$

We then obtain in (ω, k) space

$$\nabla \cdot [n(\boldsymbol{v}_{pi} + \boldsymbol{v}_{\pi i})] \approx -\mathrm{i}\frac{1}{2}nk^2\rho_i^2(\tau\omega + \omega_{*e})\frac{e\phi}{T_e} = -\mathrm{i}nk^2\rho^2(\omega - \omega_{*i})\frac{e\phi}{T_e} \tag{2.43}$$

where $\tau = T_e/T_i$ and $\rho^2 = T_e/m_i\Omega_{ci}^2$.

Equation (2.43) is in agreement with kinetic theory (compare chapter 4). The FLR effect enters as a convective contribution to the polarization drift but is in fact due to the time variation of the perturbed diamagnetic drift.

2.7.1 Effects of temperature gradients

The main source of modification in the presence of temperature gradients is the compressibility of $\boldsymbol{v}_*$. Thus equation (2.34) is changed into

$$\nabla \cdot \boldsymbol{v}_* = \frac{cn}{qB}\nabla\left(\frac{1}{n}\cdot(\hat{\boldsymbol{z}} \times \nabla T)\right). \tag{2.44}$$

Since equation (2.35) contains n only in the combination $P = nT$, equation (2.37) remains unchanged if we change the definition of κ into $\kappa_p = -(1/P_0)(\mathrm{d}P_0/\mathrm{d}x)$. We then have

$$\begin{aligned}\nabla \cdot (n\boldsymbol{v}_\pi) = &-\frac{1}{4}\rho^2\nabla n_0 \cdot \Delta_\perp \boldsymbol{v} - \frac{1}{4}\rho^2 n_0 \Delta_\perp \nabla \cdot \boldsymbol{v} + \frac{1}{4}\rho^2\kappa_p n_0 \Delta_\perp v_x \\ &-\frac{1}{4}\rho^2 n_0 \frac{\nabla T}{T}\cdot\Delta_\perp\boldsymbol{v} - \frac{1}{4m\Omega_c^2}\Delta_\perp\nabla\cdot\boldsymbol{q}_*^\perp\end{aligned} \tag{2.45}$$

where the last term is due to the $\boldsymbol{q}$ parts of equation (2.35). It is found to cancel the $\nabla \cdot \boldsymbol{v}_*$ term so that equation (2.38) is changed into

$$\nabla \cdot (n\boldsymbol{v}_\pi) = -\frac{1}{2}\rho^2\frac{1}{T}\nabla P_0 \cdot \Delta\boldsymbol{v}. \tag{2.46}$$

Since, in the presence of temperature gradients, $\boldsymbol{v}_*$ in the convective derivative of the polarization drift contains the full pressure gradient, we now find that equation (2.40) is unchanged (with our new definition of κ), and so is the conclusion in italics following it and equation (2.41). Since background pressure gradients lead in a natural way to convective pressure perturbations we now must write

$$\frac{n}{\Omega_c}\frac{\partial}{\partial t}\nabla \cdot (\hat{\boldsymbol{z}} \times \boldsymbol{v}_*) = -\frac{1}{2}\frac{\rho^2}{T}\frac{\partial}{\partial t}\Delta\delta P. \tag{2.47}$$

For the convective pressure perturbation we have

$$\frac{\delta P_i}{P} = -\frac{\omega_{*iT}}{\omega}\frac{e\phi}{T_i} \tag{2.48}$$

where ω_{*iT} is the diamagnetic drift frequency of ions due to the full background pressure gradient. Accordingly, equation (2.43) becomes

$$\nabla \cdot [n(\boldsymbol{v}_{pi} + \boldsymbol{v}_{\pi i})] = -ink^2\rho^2(\omega - \omega_{iT})\frac{e\phi}{T_e}. \tag{2.49}$$

2.8 Discussion

In this chapter we have discussed the fluid description of inhomogeneous systems and peculiarities in differences between fluid and particle drifts. These differences are important and a correct treatment is necessary for describing low frequency collective modes in inhomogeneous plasmas by fluid equations. One particular aspect of this is the description of finite Larmor radius effects, which requires the involvement of the stress tensor. This result will be compared with kinetic theory in chapter 4. We have, in this chapter, used simple Cartesian coordinates and represented magnetic field inhomogeneities by a gravity force. More complex geometries will be studied in chapter 5. We shall first continue to study new modes caused by inhomogeneities in the simple geometry in chapter 3.

2.9 Exercises

1. Assume that temperature perturbations can be neglected and derive the ratio between the perturbed diamagnetic drift and the $E \times B$ drift in the electrostatic case.
2. Express the gravity drift v_g in terms of parallel temperature for the case when $g = v_{\text{th}}^2/R_c$, i.e., g is due to a centrifugal force. Compare this drift with the usual curvature drift.
3. Verify the result that the difference between fluid and particle currents is the magnetization current. Make use of the fact that $\nabla \times \boldsymbol{e}_\parallel = \boldsymbol{e}_\parallel \times (\boldsymbol{e}_\parallel \cdot \nabla)\boldsymbol{e}_\parallel$, where $(\boldsymbol{e}_\parallel \cdot \nabla)\boldsymbol{e}_\parallel = -\boldsymbol{R}_c/R_c^2$.
4. Discuss the conditions under which the perturbed diamagnetic drift is divergence free.
5. Prove the statement after equation (2.20) that the condition $k_\perp \gg k_\parallel$ for equation (2.20) can be relaxed if we include $\boldsymbol{A}_\perp$. (Hint: use the gauge condition $\nabla \cdot \boldsymbol{A} = 0$.)
6. Go through the details in the derivation of equation (2.33).

References

[2.1] Alfvén H 1950 *Cosmical Electrodynamics* (Oxford: Oxford University Press)

[2.2] Chapman S and Cowling T G 1958 *The Mathematical Theory of Non-uniform Gases* (London: Cambridge)

[2.3] Thompson W B 1964 *An Introduction to Plasma Physics* (Oxford: Pergamon)

[2.4] Braginskii S I 1965 *Reviews of Plasma Physics* vol 1, ed M A Leontovich (New York: Consultants Bureau) p 205

[2.5] Kadomtsev B B 1965 *Plasma Turbulence* (New York: Academic)

[2.6] Mikhailovskii A B and Rudakov L I 1963 *Sov. Phys.–JETP* **17** 621

[2.7] Roberts K V and Taylor J B 1962 *Phys. Rev. Lett.* **8** 197

[2.8] Lehnert B 1964 *Dynamics of Charged Particles* (Amsterdam: North-Holland)

[2.9] Hinton F L and Horton C W 1971 *Phys. Fluids* **14** 116

[2.10] Tsai S T, Perkins F W and Stix T H 1970 *Phys. Fluids* **13** 2108

[2.11] Krall N A and Trivelpiece A W 1973 *Principles of Plasma Physics* (New York: McGraw-Hill)

[2.12] Mondt J P and Weiland J 1991 *Phys. Fluids* B **3** 3248
Mondt J P and Weiland J 1994 *Phys. Plasmas* **1** 1096

[2.13] Horton W 1984 *Handbook of Plasma Physics* vol 2, ed M N Rosenbluth and R Z Zagdeev (Amsterdam: Elsevier) pp 383–449

[2.14] Manheimer W M and Lashmore Davies C N 1989 *MHD and Microinstabilities in Confined Plasma* ed E W Laing (Bristol: Adam Hilger)

[2.15] Goldston R J and Rutherford P H 1995 *An Introduction to Plasma Physics* (Bristol: Adam Hilger)

Chapter 3

Fluid Analysis of Low Frequency Modes Driven by Inhomogeneities

We shall now apply our fluid equations discussed in chapter 2 to some fundamental modes in inhomogeneous plasmas. The literature in this field is extensive [3.1–3.49]. We shall start here by studying the effects of the inhomogeneities themselves, without complicated geometry. We also usually simplify our description to disregard temperature perturbations and background gradients. Such effects lead to considerably more complicated descriptions and are considered in chapter 5. The main reason for our interest in these modes is their potential importance for anomalous transport, and also for more macroscopic convective instabilities such as the kink instability. Since it is our intention to avoid too strong effects of geometry and boundaries we shall restrict consideration to the WKB case, i.e., $k_\perp \gg |\nabla \ln n_0|$ corresponding to large mode numbers in a torus. These modes also have $k_\parallel \ll k_\perp$, and if toroidal effects are included they require the solution of an eigenvalue problem along the magnetic field. The effects of this eigenvalue problem will only be hinted at here.

Our basic geometry will be that of a plasma slab with the density gradient in the negative x direction and the magnetic field in the z direction (figure 3.1). In a toroidal machine x corresponds to the radial coordinate, y to the poloidal coordinate and z to the toroidal coordinate. A local mode will have an extent in the radial direction which is much smaller than the typical scale of background variation. The most rapid variation, however, often takes place in the poloidal y direction, and when $k_y \gg k_x$ the equations can be conveniently simplified by neglecting k_x, as will sometimes be done in the following.

As mentioned in chapter 2, electron motion along the magnetic field lines has a stabilizing influence on the modes we consider. For small $k_\parallel$ the electron motion along the field lines is less efficient for cancelling space charge. This is the reason for our interest in modes with $k_\parallel \ll k_\perp$. In this case the main variation of the mode is in the perpendicular plane (figure 3.2). The parallel electron motion is quite different for the different modes that we shall consider in the following. Here we may separate two classes. The first class is that of drift

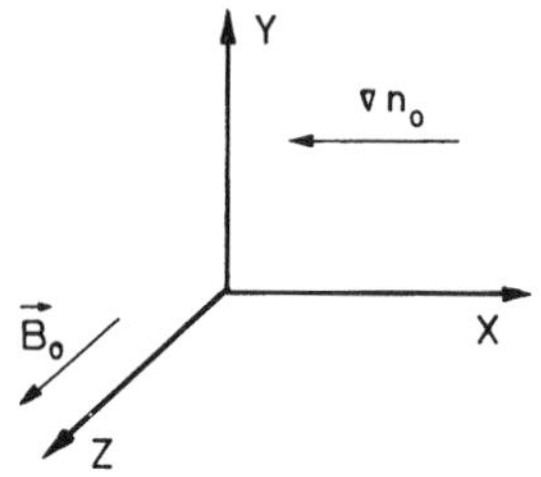

Figure 3.1. Slab picture of a magnetized plasma with a density gradient.

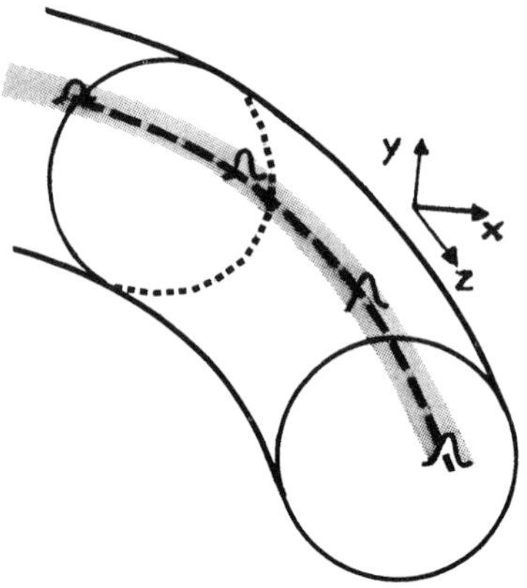

Figure 3.2. A perturbation following a field line in a torus.

waves for which $E_\parallel \neq 0$. The second class is the magnetohydrodynamic (MHD) type modes for which $E_\parallel \approx 0$. In the first case the electrons are essentially free to cancel space charge by moving along the magnetic field, while in the second case the parallel electron motion is strongly impeded either by a very small $k_\parallel$ or by electromagnetic induction. As will be shown in exercise 8, the effects of magnetic induction on $E_\parallel$ also increase for small $k_\parallel$ but the direction of propagation (sign of ω) also strongly influences $E_\parallel$, which has a maximum close to the electron diamagnetic drift frequency.

As shown in chapter 2, a vorticity $\boldsymbol{\Omega} = \nabla \times \boldsymbol{v}_E = (1/B_0)\Delta_\perp \phi \hat{\boldsymbol{z}}$ is associated with the perpendicular motion of all these modes. This means that the fluid motion forms rotating whirls. For periodic variation in x and y the velocity typically has a structure as shown in figure 3.3, where we have shown one wavelength in the y direction. The figure shows the characteristic 'smoke ring' structure caused by the opposite senses of rotation of the $E \times B$ drift around

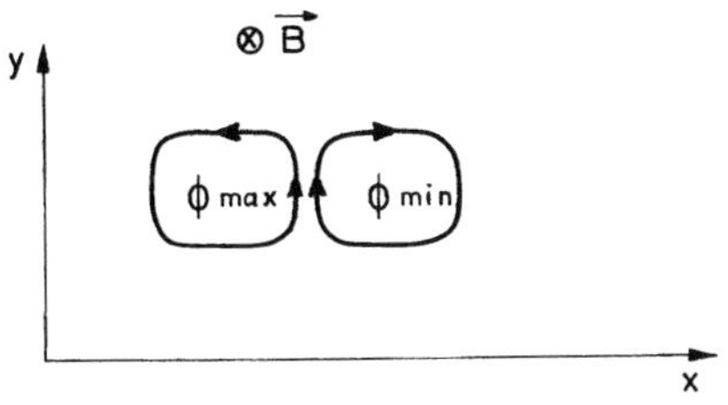

Figure 3.3. Convective cells.

potential minima and maxima. The actual fluid velocity is that shown in the figure, while the structure as such moves with the phase velocity of the wave. It is rather obvious from the figure that vortex modes are strong potential candidates for causing anomalous transport, i.e., the fluid motion (convection) tends to mix regions of higher and lower density. As is intuitively clear, however, if the perturbation is purely harmonic in time and space the fluid motion will also be completely harmonic and no net transport takes place. When there is a net damping or growth, however, this coherent picture is modified and transport takes place. This will be shown at the end of this chapter as quasi-linear diffusion.

Of particular interest in connection with convection is the convective cell mode. It has zero real part of the eigenfrequency and thus corresponds to a stationary convection in figure 3.3. In this situation a very small irreversible effect in terms of linear damping or growth or spatial 'phase mixing' is enough to cause a substantial transport.

3.1 Drift Waves

3.1.1 Elementary picture of drift waves

We now specify the background density gradient to be in the negative x direction, while the background magnetic field is in the z direction (figure 3.4). The zero order diamagnetic drift v_{*e} of the electrons due to the background density gradient will then be in the positive y direction and takes the value

$$v_{*e} = \frac{\kappa T_e}{e B_0}$$

where $\kappa = -(1/n_0)(\mathrm{d}n_0/\mathrm{d}t)$. We assume a density perturbation varying sinusoidally along y and constant along x. The variation along z is slow but fast enough to justify equation (3.2) and its linearized version equation (3.3). From equation (3.3) we shall then obtain an electric field in the y direction (figure 3.5).

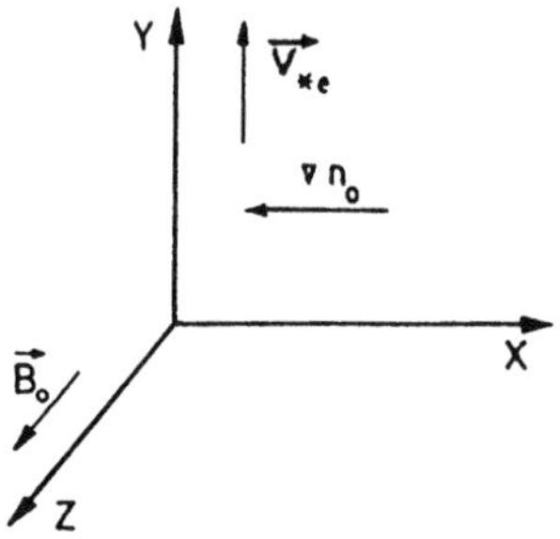

Figure 3.4. Elementary drift wave geometry.

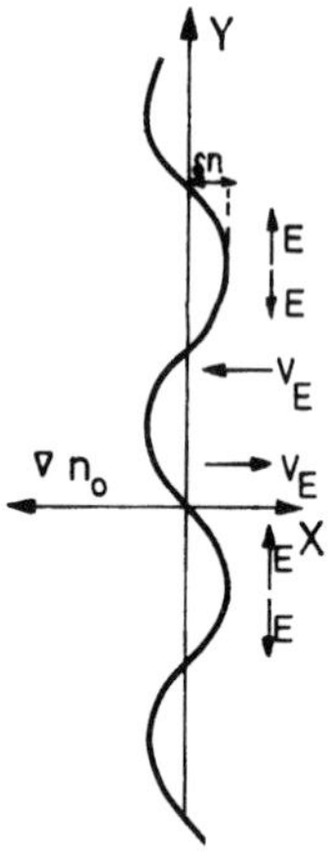

Figure 3.5. Mechanism of drift wave propagation.

In the analysis of low frequency waves, the magnitude of $k_\parallel$ is very significant. We may write the parallel equation of motion of electrons as

$$\frac{\partial v_{\parallel e}}{\partial t} + (\boldsymbol{v}_e \cdot \nabla) v_{\parallel e} = \frac{e}{m}\frac{\partial \phi}{\partial z} - \frac{1}{mn_e}\frac{\partial p_e}{\partial z} \tag{3.1}$$

where we made an electrostatic approximation and assumed the background magnetic field to be in the z direction. For low frequency processes we may drop the electron inertia terms on the left-hand side of equation (3.1). For isothermal electrons, $\partial p_e/\partial z = T_e(\partial n_e/\partial z)$, and we then have

$$e\frac{\partial \phi}{\partial z} - \frac{T_e}{n_e}\frac{\partial n_e}{\partial z} = 0$$

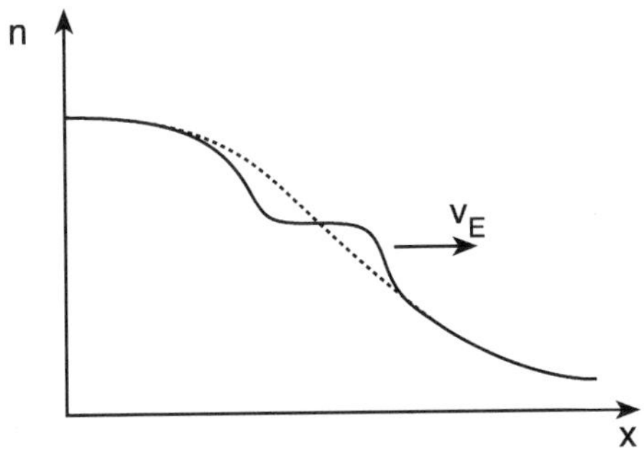

Figure 3.6. Convective density perturbation.

or in integrated form

$$\frac{n_e}{n_0} = \mathrm{e}^{e\phi/T_e} \tag{3.2}$$

where we introduced the equilibrium value n_0 of n_e. Equation (3.3) is the Boltzmann distribution, which is usually a good approximation if $k_\parallel$ is not too small. Writing $n_e = n_0 + \delta n_e$ and expanding the exponential for $e\phi/T_e$ we find

$$\frac{\delta n_e}{n_0} = \frac{e\phi}{T_e}. \tag{3.3}$$

This result is in agreement with our previous estimate for the validity of quasi-neutrality. This field will cause an $E \times B$ drift v_{Ex} in the x direction as shown in figure 3.6. This drift will, due to the background density gradient, cause a change of density in such a way that the perturbation moves in the positive y direction (figure 3.5). Since $\nabla \cdot (n\boldsymbol{v}_{*j}) = 0$ and $\nabla \cdot \boldsymbol{v}_E = 0$, we obtain from the linearized continuity equation for ions by dropping $\boldsymbol{v}_{pi}$, $\boldsymbol{v}_{\pi i}$, $\boldsymbol{v}_{gi}$ and $\boldsymbol{v}_{\parallel i}$

$$\frac{\partial n_i}{\partial t} + v_{Ex}\frac{\mathrm{d}n_0}{\mathrm{d}x} = 0. \tag{3.4}$$

In equation (3.4) we dropped the parallel motion of the ions. This is permitted for small enough $k_\parallel$ since $\nabla \cdot (n\boldsymbol{v}_\parallel) \approx \mathrm{i}nk_\parallel v_\parallel$. Equation (3.4) describes the density variation due to convection mentioned above. It corresponds to an incompressible motion ($\nabla \cdot \boldsymbol{v} = 0$). Introducing

$$v_{Ex} = -\frac{1}{B_0}\frac{\partial\phi}{\partial y}$$

we have from equation (3.4)

$$\frac{1}{n_0}\frac{\partial n_i}{\partial t} + \frac{1}{B_0}\frac{\partial\phi}{\partial y} = 0.$$

By using the quasi-neutrality condition and equation (3.3) we now have

$$\frac{\partial \phi}{\partial t} + v_{*e}\frac{\partial \phi}{\partial y} = 0$$

corresponding to the dispersion relation

$$\frac{\omega}{k_y} = v_{*e}. \tag{3.5}$$

Equation (3.5) shows that the velocity of propagation of the density perturbation in figure 3.5 is the electron diamagnetic drift velocity v_{*e}. This is the simplest form of a drift wave.

3.1.2 Effects of finite ion inertia

We are now interested in extending the results of the previous section. As it turns out, the effects of ion inertia, which cause the drift motion of electrons and ions to be different, are also associated with compressibility. First, we note that the Boltzmann distribution for the electron density in equation (3.3) is also obtained for ion acoustic waves propagating along B_0 and corresponds to an expansion of the kinetic integral for the density perturbation in the limit $\omega/k_\parallel \ll v_{\text{th}\,e}$. If we assume the ion temperature to be very small, so that the region

$$v_{\text{th}\,i} \ll \frac{\omega}{k_\parallel} \ll v_{\text{th}\,e} \tag{3.6}$$

usually considered for drift waves is wide, we may drop the ion pressure term and obtain

$$v_{\parallel i} = \frac{k_\parallel}{\omega}\frac{e\phi}{m_i}. \tag{3.7}$$

Including the ion polarization drift for the perpendicular motion, and still assuming $k_x = 0$ we have

$$\boldsymbol{v}_{\perp i} = \mathrm{i}k_y\frac{\phi}{B_0}\hat{\boldsymbol{x}} - k_y\frac{\phi}{\Omega_{ci}}\hat{\boldsymbol{y}} + \boldsymbol{v}_{*i}. \tag{3.8}$$

Introducing equation (3.8) into the ion continuity equation, assuming $k_y \gg \kappa$ so that $\nabla(n_0\boldsymbol{v}_p) = n_0\nabla\cdot\boldsymbol{v}_p$, using equation (2.8) and $\nabla\cdot\boldsymbol{v}_E = 0$ we find

$$\frac{\delta n_i}{n_0} = \left(\frac{\omega_{*e}}{\omega} - \frac{k_y^2 T_e}{m_i\Omega_{ci}^2} + \frac{T_e}{m_i}\frac{k_\parallel^2}{\omega^2}\right)\frac{e\phi}{T_e} \tag{3.9}$$

where we introduced the drift frequency

$$\omega_{*e} = k_y v_{*e}. \tag{3.10}$$

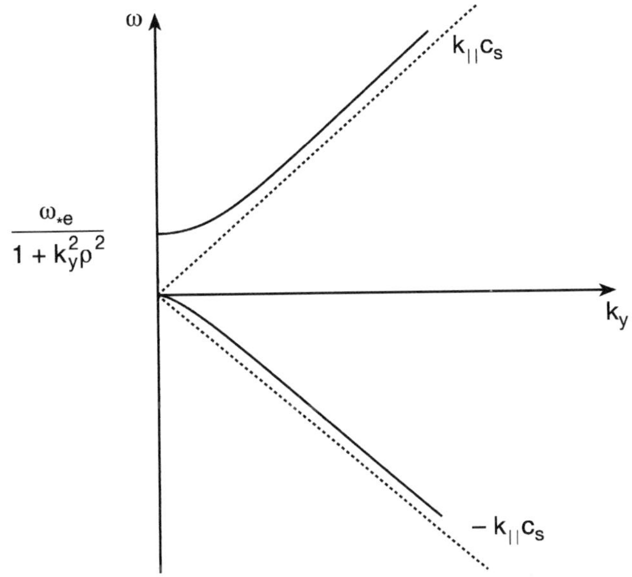

Figure 3.7. Two-dimensional dispersion diagram of drift waves.

Combining equation (3.9) with equation (3.3) and using the quasi-neutrality condition we obtain the dispersion relation

$$\omega^2(1 + k_y^2\rho^2) - \omega\omega_{*e} - k_\parallel^2 c_s^2 = 0 \tag{3.11}$$

where we introduced $\rho = c_s/\Omega_{ci}$ and $c_s = (T_e/m_i)^{1/2}$. The term $k_y^2\rho^2$ originates from the ion polarization drift and represents the influence of ion inertia. It enters as a Larmor radius effect, where ρ is the ion Larmor radius at the electron temperature. The dispersion relation (3.11) is represented in figure 3.7.

From figure 3.7 we realize that for large $k_\parallel$ the drift wave turns into the ion acoustic wave. Clearly ion parallel motion may be neglected when $k_\parallel c_s \ll \omega_{*e} = k_y v_{*e}$ (if $k_y\rho \ll 1$). For comparison we note that for typical JET parameters we have $c_s \approx 10^6$ m s^{-1} and $v_{*e} \approx 10^3$ m s^{-1}, i.e., we have to require $k_\parallel \ll k_y \times 10^{-3}$ in order to drop the parallel ion motion. For the Boltzmann distribution of electrons to be valid we require $k_\parallel v_{\text{th}\,e} \gg k_y v_{*e}$, which means $k_\parallel \gg 0.25 \times 10^{-4} k_y$ for JET. The interest in such small $k_\parallel$ is mainly due to the fact that instability is likely to occur in this region (compare figure 2.5).

3.1.3 Drift instability

As long as the electrons are free to move along B_0 to cancel space charge, the Boltzmann relation (3.3) is fulfilled and the drift wave is stable. There are, however, several effects that may limit the mobility of the electrons so as to modify equation (3.3). These effects are generally more important for small $k_\parallel$, e.g., electron–ion collisions, Landau damping, electron inertia or inductance.

If the electrons are not able to move completely freely a phase shift will appear, corresponding to a time lag between density and potential in equation (3.3). We then modify equation (3.3) as

$$\frac{\delta n_e}{n_0} = \frac{e\phi}{T}(1 - \mathrm{i}\delta). \tag{3.12}$$

By replacing equation (3.3) with equation (3.12) in the derivation of equation (3.5) we obtain the result

$$\omega = \frac{k_y v_{*e}}{1 - \mathrm{i}\delta} \approx k_y v_{*e}(1 + \mathrm{i}\delta) \tag{3.13}$$

if we assume $\delta \ll 1$. We note that due to the time variation $\exp(-\mathrm{i}\omega t)$, $\delta > 0$ means that the potential lags behind the density. This situation corresponds to an instability.

3.1.4 Excitation by electron–ion collisions

As an example we shall now consider the collisional drift instability. We assume the ordering

$$\omega \ll \nu_{ei} \ll \Omega_{ci} \tag{3.14}$$

where ν_{ei} is the electron–ion collision frequency. As a result of equation (3.14) we may for small $k_\parallel$ include the effect of electron–ion collisions on the electron parallel motion, but continue to drop electron inertia. Dropping the parallel ion motion we then find

$$v_{\parallel e} \approx \mathrm{i}\frac{k_\parallel T_e}{\nu_{ei} m_e}\left(\frac{e\phi}{T_e} - \frac{\delta n_e}{n_0}\right). \tag{3.15}$$

Taking the limit $k_x = 0$ we have

$$\boldsymbol{v}_{\perp e} = -\mathrm{i}k_y \frac{\phi}{B_0}\hat{\boldsymbol{x}} + \boldsymbol{v}_{*e}.$$

The electron continuity equation now yields

$$-\mathrm{i}\omega\frac{\delta n_e}{n_0} + \mathrm{i}k_y \frac{\kappa}{B_0}\frac{T_e}{e}\frac{e\phi}{T_e} - \frac{k_\parallel^2}{\nu_{ei}}\frac{T_e}{m_e}\left(\frac{e\phi}{T_e} - \frac{\delta n_e}{n_0}\right) = 0$$

which may be reduced to

$$\frac{\delta n_e}{n_0} \approx \frac{e\phi}{T_e}\frac{\omega_{*e} + \mathrm{i}k_\parallel^2 D_\parallel}{\omega + \mathrm{i}k_\parallel^2 D_\parallel} \tag{3.16}$$

where

$$D_{\parallel} = T_e/m_e\nu_{ei} \tag{3.17}$$

is the parallel diffusion coefficient.

For the orderings already introduced it is reasonable to assume that $k_{\parallel}^2 D_{\parallel} \gg \omega$.

Thus expanding equation (3.16) we obtain

$$\frac{\delta n_e}{n_0} = \frac{e\phi}{T_e}\left[1 - \mathrm{i}\frac{m_e\nu_{ei}}{k_{\parallel}^2 T_e}(\omega_{*e} - \omega)\right]. \tag{3.18}$$

By identifying δ in equation (3.12) with the corresponding expression in equation (3.18) we find from equation (3.13) that we have an instability if $\omega < \omega_{*e}$. We find

$$\operatorname{Im}\omega = \frac{m_e\nu_{ei}}{k_{\parallel}^2 T_e}\omega_{*e}(\omega_{*e} - \omega). \tag{3.19}$$

Since, however, we have $\operatorname{Re}\omega = \omega_{*e}$, from equation (3.13) we realize that we need some additional effect in order to have an instability. Since the ions are not so strongly influenced by collisions with electrons we use equation (3.9) with $k_{\parallel} = 0$ for the ion density perturbation. Combining this equation with equation (3.18) we find the dispersion relation

$$\omega(1 + k_y^2\rho^2) = \omega_{*e} + \nu_{ei}\frac{m_e}{k_{\parallel}^2 T_e}\omega(\omega_{*e} - \omega).$$

We now assume that $k_y^2\rho^2 \ll 1$. We note also that the last term is small due to the assumption $\alpha \ll \omega$. Thus dividing by $1 + k_y^2\rho^2$, expanding this denominator in the ω_{*e} term and dropping $k_y^2\rho^2$ in the last term we have

$$\omega = \omega_{*e}(1 - k_y^2\rho^2) + \mathrm{i}\nu_{ei}\frac{m_e}{k_{\parallel}^2 T_e}\omega(\omega_{*e} - \omega). \tag{3.20}$$

Writing the solution as $\omega = \omega_r + \mathrm{i}\gamma$ where $\gamma \ll \omega_r$ we find

$$\omega_r \approx \omega_{*e}(1 - k_y^2\rho^2) \tag{3.21}$$

and

$$\gamma \approx \frac{\nu_{ei} m_e}{k_{\parallel}^2 T_e}\omega_{*e}^2 k_y^2\rho^2. \tag{3.22}$$

We see from equation (3.22) that the ion inertia $k_y^2\rho^2$ is essential for an instability to develop. We may explain the instability in the way that the ion inertia causes the particle drifts of electrons and ions in the perpendicular plane to become different. This leads to charge separation effects if we have a density perturbation, and as a result of the electron–ion collision the electrons are not able to instantly neutralize the charge separation by moving along the magnetic field.

3.2 MHD-Type Modes

As mentioned at the beginning of this chapter there are two classes of low frequency modes: drift modes and MHD-type modes. While the drift modes are characterized by essentially free electron motion along the field lines leading to the Boltzmann distribution in equation (3.3) or minor modifications thereof, the MHD-type modes are modes where the parallel electric field to lowest order vanishes. This can, in the electrostatic case, be accomplished by a very small $k_\parallel$ ($\omega \gg k_\parallel v_{\mathrm{th}\,e}$) and in the electromagnetic case by a cancellation between electrostatic and induction parts of E. In both cases the parallel electron motion is strongly impeded and, as a consequence, new types of instability may arise. These unstable modes may be divided into two classes: *pressure-driven modes*, here represented by interchange and ballooning modes, and *current-driven modes*, here represented by kink modes. The transition between MHD and drift-type modes in a simple case is shown by exercise 8.

3.2.1 Interchange modes

One of the most dangerous modes in fusion machines is the interchange mode, sometimes also called the flute mode, which tends to interchange 'flux tubes' of different pressure, thus causing a convective transport. (Compare also section 5.4.) Interchange modes are unstable when the magnetic curvature generates a centrifugal force, due to thermal motion along the field lines, which is directed in the opposite direction to the pressure gradient. As a simple example we consider a z-pinch with poloidal magnetic field only. Figure 3.8 shows fluid elements that would tend to change places.

A simple fluid analogue of this instability is the Rayleigh–Taylor instability (figure 3.9) when a heavy fluid is resting on a light fluid. The density gradient corresponds here to the pressure gradient for the interchange mode, while the gravity represents the centrifugal force.

The gravity may thus be used to simulate a curvature and this is the main reason why we included it in equation (2.1). We shall here neglect finite Larmor radius effects that would correspond to diamagnetic drift contributions to the polarization drift and stress tensor drifts. We shall also make the approximation $k_\parallel = 0$ (flute mode). This is the most unstable mode since a mode with $k_\parallel \neq 0$ would tend to bend the frozen in magnetic field lines, thus increasing the magnetic energy. We may obtain a dispersion relation by substituting the drifts into the low frequency condition

$$\nabla \cdot j = 0.$$

In the present case this gives

$$\nabla \cdot [en(\boldsymbol{v}_{pi} + \boldsymbol{v}_{gi} - \boldsymbol{v}_{ge})] = 0. \tag{3.23}$$

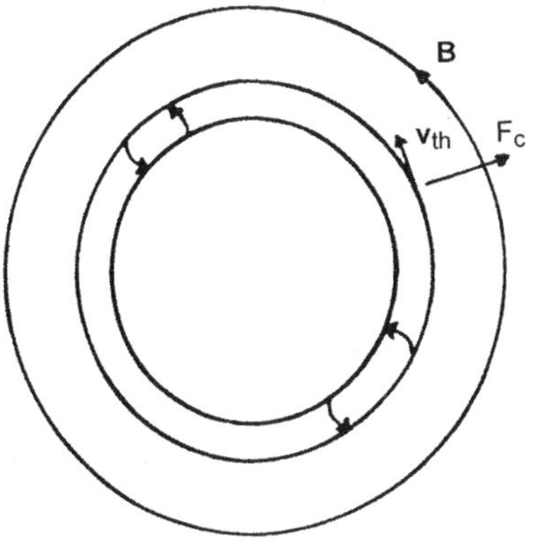

Figure 3.8. Interchange of fluid elements in a cylinder.

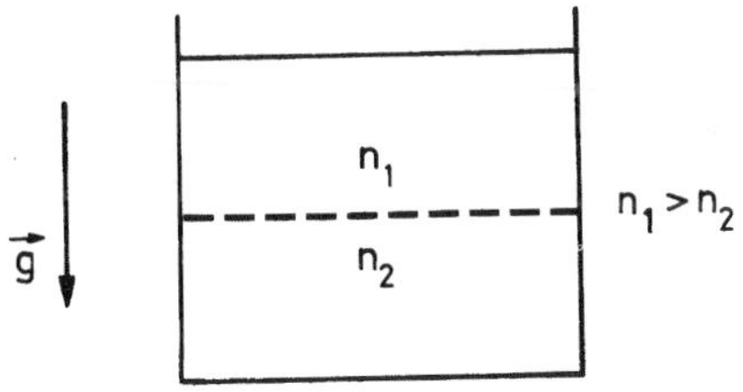

Figure 3.9. Rayleigh–Taylor instability.

A linearization, again using $k_\perp \gg \kappa$, leads to

$$n_0 \nabla \cdot \left[-\frac{1}{B_0 \Omega_{ci}} \left(\frac{\partial}{\partial t} + \boldsymbol{v}_{gi} \cdot \nabla \right) \nabla_\perp \phi \right] + (\boldsymbol{v}_{gi} - \boldsymbol{v}_{ge}) \cdot \nabla \delta n = 0. \qquad (3.24)$$

The density perturbation can be obtained from the electron continuity equation. We consider $\boldsymbol{v}_{ge}$ to be small. We may then ignore it in the continuity equation

$$\frac{\partial \delta n_e}{\partial t} + \boldsymbol{v}_E \cdot \nabla n_0 = 0$$

or in $(\boldsymbol{k}, \omega)$ space

$$\frac{\delta n_e}{n_0} = \frac{\omega_{*e}}{\omega} \frac{e\phi}{T_e}. \qquad (3.25)$$

Substituting equation (3.25) into equation (3.24) we obtain the dispersion relation

$$\omega(\omega - k_y v_{gi}) = -\kappa \left(g_i + \frac{m_e}{m_i} g_e \right) \frac{k_y^2}{k_\perp^2} \qquad (3.26)$$

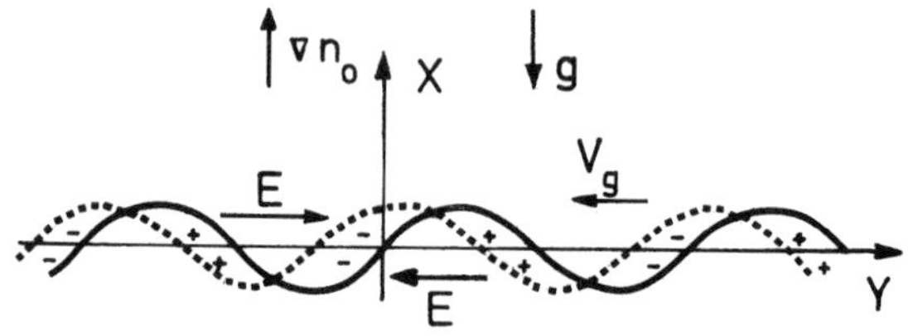

Figure 3.10. The mechanism of interchange instability.

where $\kappa = -\mathrm{d}\ln n/\mathrm{d}x$. We note here that if the gravity is replaced by a real curvature $\kappa \rightarrow -\mathrm{d}\ln P/\mathrm{d}x$. We easily recognize here the part $\omega^2 = -\kappa g$, corresponding to the Rayleigh–Taylor instability. When g_j is due to curvature we have $g_j = 2T_j/m_j R_c$, where R_c is the radius of curvature. Then the dispersion relation may be rewritten as

$$\omega(\omega - k_y v_{gi}) = -\frac{2\kappa(T_e + T_i)}{m_i R_c}\frac{k_y^2}{k_\perp^2}. \tag{3.27}$$

The drift $k_y v_{gi}$ here is stabilizing. This means that modes with small k_y are the most unstable modes. As it turns out, the lowest order FLR correction has the same influence but is typically larger than the drift term preserved here. In the fluid description this instability is related to the density gradient in a very simple way, i.e., the fluid motion happens in the direction of the gradient. In a plasma the convection is also in the direction of the pressure gradient but the actual physical process is more complicated, since in a plasma forces in the perpendicular plane primarily give rise to motion perpendicular to the force. The source of the instability is the difference in gravity (curvature) drifts of electrons and ions, which in combination with a density perturbation leads to a charge separation. When $k_\parallel = 0$ the electrons cannot short circuit this charge separation, which leads to an electric field that is perpendicular to the pressure gradient and the magnetic field. When the pressure gradient and the gravity have opposite directions this electric field causes an $E \times B$ drift, which enhances the original perturbation (figure 3.10). The real frequency caused by $k_y v_{gi}$ is stabilizing since it changes the polarity of the field. A gravity due to field line bending is shown in figure 3.11.

3.2.2 The convective cell mode

If we let $\kappa \rightarrow 0$ in equation (3.27) the two branches become uncoupled and we have one mode with $\omega = 0$ and one with $\omega = k_y v_{gi}$. The mode $\omega = 0$ corresponds to a stationary convection (compare figure 3.3) and is called the convective cell mode. When finite ion Larmor radius effects are included, the ion diamagnetic drift frequency is added to v_{gi}. This means that we also have two different modes if we let $R_c \to \infty$. In a real system with curvature and density gradient the convective cell thus turns into the interchange mode. Since it has no variation along the field lines it will experience the average curvature along the field line. This curvature is usually favourable in a tokamak leading to a real eigenfrequency. This will tend to reduce the transport (compare chapter 7).

3.2.3 Electromagnetic interchange modes

In a physical system with magnetic shear (see section 5.2) the approximation $k_\parallel = 0$ cannot be exactly fulfilled since the mode has a finite extension in space, and since the magnetic field direction is space dependent. Another situation may be when the average curvature is stabilizing but there are local regions along a field line where the curvature is destabilizing. In order to see qualitatively what the consequences of a finite $k_\parallel$ will be, we shall here simply include a finite $k_\parallel$ in our simple slab geometry. In this case our previous description for the perpendicular motion continues to hold. For the parallel direction we may neglect ion motion assuming $\omega \gg k_\parallel c_s$ (compare equation (3.11)). The parallel electron motion can in the simplest case be described in the same way as for shear Alfvén waves, i.e., we combine the Ampère law along the field lines (2.18) with the perfect conductivity condition (2.19). Then using equation (2.17) in the form

$$\nabla \cdot \boldsymbol{j}_\parallel = -\nabla_\perp \cdot \boldsymbol{j}_\perp$$

we arrive at the dispersion relation

$$\omega(\omega - k_y v_{gi}) - k_\parallel^2 v_A^2 + \frac{2\kappa(T_e + T_i)}{m_i R_c}\frac{k_y^2}{k_\perp^2} = 0. \tag{3.28}$$

The new effect here is the bending of the field lines, represented by the Alfvén frequency. This effect is stabilizing since the line bending increases the magnetic energy (figure 3.12). For small k_y the dispersion relation (3.28) leads to a pressure balance condition for stability. In a torus with periodic curvature and unfavourable curvature regions of length $L_c \approx 2\pi q R_c$, where q is the safety factor (compare the chapter on toroidal mode structure), we may to order of magnitude take $k_\parallel \approx 1/qR$, where R is the large radius and $\kappa \approx 1/a$, where a is the small radius. The pressure balance condition for stability then takes the simple form

$$\beta < \beta_c = \frac{a}{Rq^2} \tag{3.29}$$

where $\beta = 2\mu_0 n(T_i + T_e)/B^2$ is the ratio of plasma and magnetic field pressure. This β limit is typical of ballooning modes in toroidal machines. These modes are interchange modes localized in regions of unfavourable curvature and are believed to limit the achievable β in tokamaks (figure 3.14).

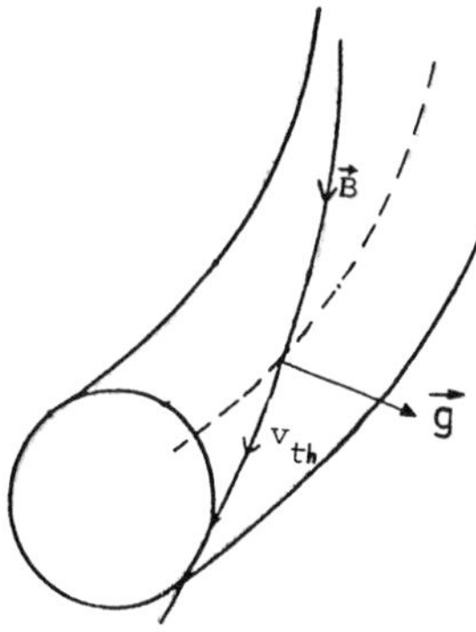

Figure 3.11. Gravity representing a centrifugal force.

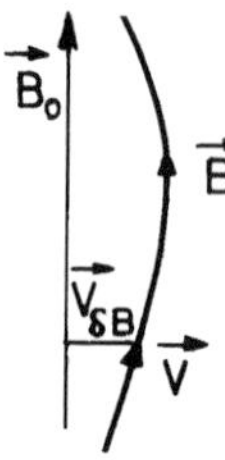

Figure 3.12. The magnetic energy increases when the field line bends. This effect is stabilizing.

Another source of finite $k_\parallel$ is the radial extent of a mode in a system with shear. We may here think of the previous slab quantities as averaged over the mode profile. The average curvature may be written as

$$\left\langle \frac{\kappa}{R_c} \right\rangle \approx -\frac{\delta}{R}\frac{1}{p}\frac{\mathrm{d}P}{\mathrm{d}r}$$

where δ is a factor due to averaging and we introduced the radial coordinate r instead of x.

The averaging of $k_\parallel$ leads to (compare chapter 5)

$$\langle k_\parallel \rangle \approx \Delta' s/qR$$

where $s = \mathrm{d}\ln q/\mathrm{d}\ln r$ and r is the radial coordinate. We then obtain the stability condition

$$\Delta'^2 s^2 > -\delta\beta q^2 R \frac{\mathrm{d}\ln p}{\mathrm{d}r}.$$

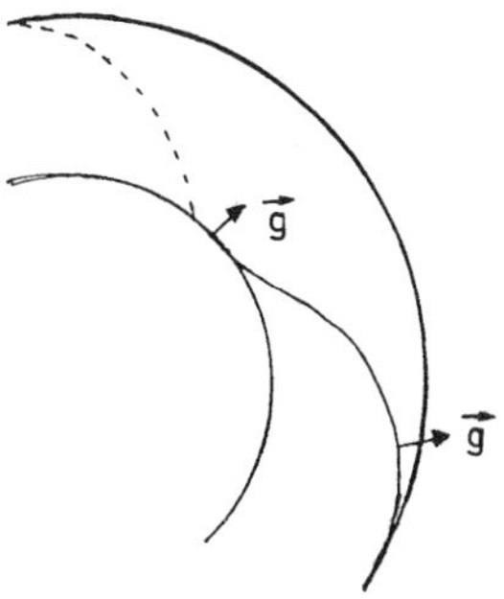

Figure 3.13. Field line curvature.

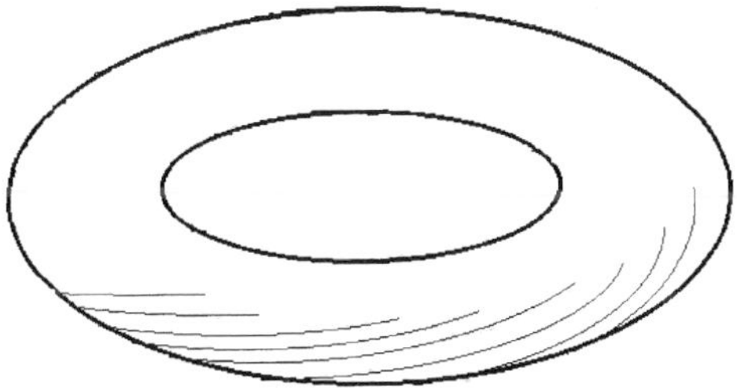

Figure 3.14. Ballooning mode perturbations on a torus.

As it turns out (a discussion of this result will be given later)

$$\Delta' \approx \tfrac{1}{2}. \tag{3.30}$$

For a z-pinch where the magnetic field is purely poloidal the problem becomes singular since $q = 0$. If, however, we treat q as small but finite we obtain $\delta = -r/Rq^2$ and the q dependence disappears. We then obtain the *Suydam criterion* [3.1]:

$$\frac{1}{4}s^2 + r\frac{\mathrm{d}\beta}{\mathrm{d}r} > 0. \tag{3.31}$$

For a torus with both poloidal and toroidal magnetic field it can be shown that $\delta = (r/R)(1 - 1/q^2)$ and the stability condition turns into the *Mercier criterion* [3.2]:

$$\frac{1}{4}s^2 + r\frac{\mathrm{d}\beta}{\mathrm{d}r}(1 - q^2) > 0. \tag{3.32}$$

This means that the average curvature is stabilizing for $q > 1$ and instability is only possible for $q < 1$.

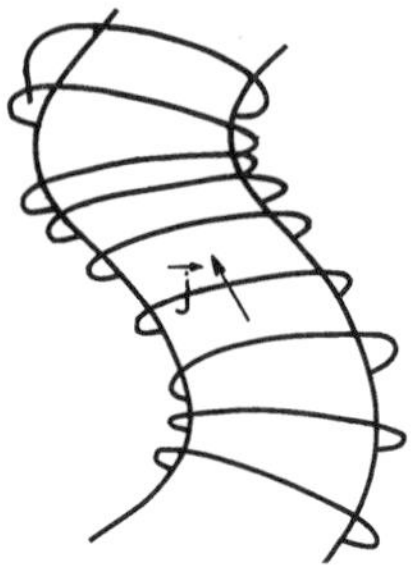

Figure 3.15. Kink perturbation.

The Mercier criterion only holds for modes that are highly elongated along the field lines and experience only the average curvature (figure 3.13). When localized ballooning modes are taken into account the possibility for instability increases strongly and a rather typical β limit is $\beta_c/2$.

3.2.4 Kink modes

One of the most dangerous instabilities in current-carrying cylindrical and toroidal plasmas is the kink instability. It corresponds to a bending of the whole system (global mode) so that the change in magnetic pressure tends to increase the perturbation (see figure 3.15).

Although this mode is most easily visualized for global perturbations in combination with sharp current boundaries, the only necessary ingredient is a background current gradient perpendicular to the magnetic field.

We shall here include the kink mode in our previous analysis by including a background current with a gradient in the x (radial) direction. If we neglect the associated frequency shift ($\omega \gg k_\parallel v_{\parallel 0}$) the only new terms we need to include are the $\boldsymbol{v}_{\delta B}$ drifts in equation (2.10) for electrons and ions. This leads to a new contribution to $\nabla_\perp \cdot \boldsymbol{j}$ as

$$\nabla \cdot \left[e(n_i \boldsymbol{v}_{\parallel 0i} - n_e \boldsymbol{v}_{\parallel 0e}) \frac{\delta \boldsymbol{B}_\perp}{B_\parallel} \right] = \frac{\delta \boldsymbol{B}_\perp}{B_\parallel} \cdot \nabla \boldsymbol{J}_{\parallel 0} \tag{3.33}$$

where $\boldsymbol{J}_{\parallel 0}$ is the background current.

Keeping also our previous driving pressure term we may write the equation $\nabla \cdot \boldsymbol{j} = 0$ as

$$-\frac{1}{\mu_0} \hat{\boldsymbol{z}} \cdot \nabla \Delta_\perp A_\parallel = -\nabla_\perp \cdot \left[en\boldsymbol{v}_{pi} + en(\boldsymbol{v}_{gi} - \boldsymbol{v}_{ge}) + J_{\parallel 0} \frac{\delta \boldsymbol{B}_\perp}{B} \right] \tag{3.34}$$

where $A_\parallel = -(\mathrm{i}/\omega)\hat{\boldsymbol{z}} \cdot \nabla \phi$ according to the condition that $E_\parallel = 0$. Proceeding as before, but now also neglecting for simplicity the frequency shift due to the ion gravity drift, we obtain the dispersion relation for $k_\perp \gg |\nabla \ln n|$

$$\omega^2 - k_\parallel^2 v_A^2 + 2\frac{\kappa(T_e + T_i)}{m_i R_c}\frac{k_y^2}{k_\perp^2} - \frac{B_0}{n_0 m_i}\frac{k_\parallel k_y}{k_\perp^2}\frac{\mathrm{d}J_{\parallel 0}}{\mathrm{d}x} = 0. \tag{3.35}$$

Since $\nabla \cdot \delta \boldsymbol{B}_\perp = 0$ the $\nabla \cdot$ has to operate on the background quantity $J_{\parallel 0}$ in the last term of equation (3.35).

This has the consequence that the kink term is usually small for local modes since also $k_\parallel$ is small. As it turns out, however, the only new term that arises if we relax the condition $k_\perp \gg |\nabla \ln n_0|$, $|\nabla \ln J_{\parallel 0}|$ is the density gradient contribution from the polarization drift. This effect is usually neglected for global modes so that equation (3.35) can in fact be written as an eigenvalue equation for such modes. The kink term may also become important locally if the background current gradient is locally large. With $\nabla \ln n_0 \approx -1/a$, where a is the small radius, the β limit in equation (3.29) is modified to

$$\beta \leq \frac{a}{Rq^2} + \frac{\mu_0 a}{B_0 q k_\perp}\frac{\mathrm{d}J_{\parallel 0}}{\mathrm{d}x} \tag{3.36}$$

where we used $k_\parallel \approx 1/qR$. The condition (3.36) shows that the kink term decreases the β limit (destabilizing) if $\mathrm{d}J_{\parallel 0}/\mathrm{d}x < 0$, which is the typical case. If we write $\mathrm{d}J_{\parallel 0}/\mathrm{d}x = -\kappa_b e n v_{\parallel 0}$, we obtain the β limit in the form

$$\beta \leq \frac{a}{Rq^2} - \frac{\Omega_{ci} v_{\parallel 0}}{k_\perp v_A^2}\kappa_b a. \tag{3.37}$$

For the kink term to change the β limit appreciably we need $\kappa_b a = 5$ for typical tokamak parameters if $k_\perp v_A \sim \Omega_{ci}$.

3.2.5 Stabilization of electrostatic interchange modes by parallel electron motion

In the preceding section we found that interchange modes may be unstable for zero $k_\parallel$ or when $\kappa g > k_\parallel^2 v_A^2$. In the opposite case when $k_\parallel v_A$ becomes large the mode becomes electrostatic. (This will come out of the kinetic treatment in chapter 4 but can also easily be seen from the fluid equations, as shown by exercise 4.)

In the electrostatic limit when $\omega \ll k_\parallel v_{\mathrm{th}\,e}$ the electrons are Boltzmann-distributed, i.e.

$$\frac{\delta n_e}{n} = \frac{e\phi}{T_e}. \tag{3.38}$$

This relation comes out of the parallel equation of motion and is not influenced by gravitation. For the ions we may use equation (3.9), where the gravitation introduces a Doppler shift, i.e., $\omega \to \omega - k_y v_{gi}$, and where we neglect $k_\parallel^2 c_s^2$

$$\frac{\delta n_i}{n} = \left(\frac{\omega_{*e}}{\omega - k_y v_{gi}} - k_y^2 \rho^2\right)\frac{e\phi}{T_e}. \tag{3.39}$$

Then, using quasi-neutrality we obtain the dispersion relation

$$\omega = \frac{\omega_{*e}}{1 + k_y^2 \rho^2} + k_y v_{gi}. \tag{3.40}$$

This is just the dispersion relation for an ordinary drift wave where a frequency shift $k_y v_{gi}$ has been added. Thus there is no instability. The reason for this is that the electrons are free to move along the field lines (compare figure 2.5) to cancel space charge. In the electromagnetic case the electron motion along the field lines is impeded by magnetic induction, thus providing the necessary conditions for instability.

3.2.6 FLR stabilization of interchange modes

As it turns out, the lowest order FLR effect is often significant and introduces qualitatively new effects, while the higher order effect usually only modifies previously known results. We shall here demonstrate the stabilizing influence of FLR effects on interchange modes. This is now very easily done by just replacing $\nabla \cdot (n\boldsymbol{v}_{pi})$ with

$$\nabla \cdot [n(\boldsymbol{v}_{pi} + \boldsymbol{v}_{\pi i})] = ink^2\rho^2(\omega - \omega_{*i})\frac{e\phi}{T_e}.$$

This result can be verified by using the stress tensor $\boldsymbol{\pi}$. This means that the lowest order FLR term can be obtained by just shifting ω by ω_{*i} in the ion polarization drift. This effect is typically larger than the shift due to the gravity drift in equation (3.26), as can be seen from the following estimate. When v_{gi} is due to curvature it may be written

$$|v_{gi}| = \frac{2T_i}{mR_c\Omega_{ci}} = \frac{\rho_i}{R_c} v_{\mathrm{th}\,i}. \tag{3.41}$$

The diamagnetic drift may be written

$$|v_{*i}| \approx \frac{T_i \kappa}{mR_c\Omega_{ci}} \approx \frac{1}{2}\kappa\rho_i v_{\mathrm{th}\,i}. \tag{3.42}$$

Now R_c is typically approximated by the large radius R in a torus and κ is typically $1/a$, where a is the small radius. We thus arrive at the estimate

$$\left|\frac{v_{gi}}{v_{*i}}\right| \approx 2\frac{a}{R}. \tag{3.43}$$

When $a \ll R$ we may thus neglect the gravity drift (this is no longer fulfilled for the newest generation of large tokamaks).

The dispersion relation then takes the form

$$\omega(\omega - \omega_{*i}) - k_\parallel^2 v_A^2 + \frac{2\kappa(T_e + T_i)}{m_i R}\frac{k_y^2}{k_\perp^2} = 0 \tag{3.44}$$

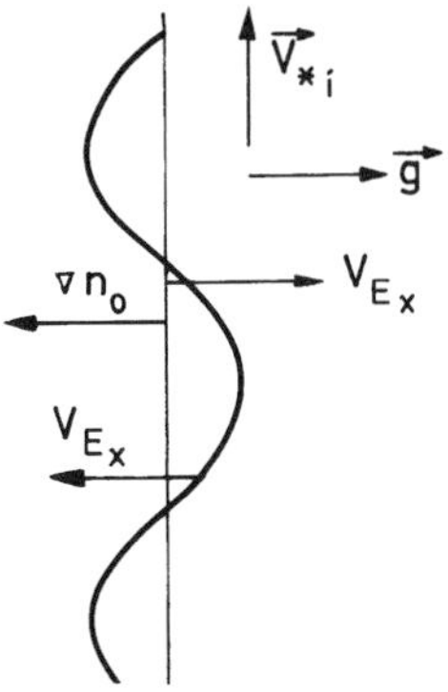

Figure 3.16. The ion diamagnetic drift causes a propagation which changes the polarity of the charge separation.

with the solution

$$\omega = \frac{1}{2}\omega_{*i} \pm \sqrt{\frac{1}{4}\omega_{*i}^2 + k_\parallel^2 v_A^2 - 2\kappa \frac{T_e + T_i}{m_i R} \frac{k_y^2}{k_\perp^2}}. \tag{3.45}$$

The stabilizing FLR term changes the β limit as given by equation (3.29) to

$$\beta < \frac{a}{q^2 R} + \frac{1}{4}\omega_{*i}^2 aR/v_A^2. \tag{3.46}$$

This is an important effect for modes with large k_y. In a tokamak there are, however, unstable modes with such small k_y that FLR effects can be neglected within the present description. When curvature effects that are higher order in a/R and geometrical effects in D-shaped tokamaks are included, however, it appears that finite Larmor radius effects have to be included as well. In addition to the stabilizing effect we also notice from equation (3.45) that the mode now has a finite real part of the eigenfrequency also at marginal stability and in the unstable case. This real part of ω corresponds to propagation in the y direction and is in fact the reason for the stabilizing effect.

The interchange of fluid elements by convection, which is the fundamental instability we are considering, is a fluid motion and the fluid only propagates in the x direction (if we neglect k_x). The perturbation, however, propagates in the y direction, causing a change in the direction of convection after a time $\tau = 2\pi/\omega_{*i}$ (figure 3.16). We then realize that if this time is short, as compared to the time needed for the instability to develop, it will become stabilized. The physical interpretation of the FLR effect is a modification of the $E \times B$ drift due to the inhomogeneity of $\boldsymbol{E}$ along the gyro orbit. This effect is usually only important for ions and leads to a charge separation in the presence of a density gradient. This charge separation will have a different sign in the region where v_E is directed in the positive or negative x direction, and leads to a propagation of the perturbation in the y direction.

3.2.7 Kinetic Alfvén waves

In order to exemplify the modification of MHD-type modes in the presence of a finite $E_\parallel$, we shall now study a mode which is of a particular interest for the heating of fusion plasmas: the kinetic Alfvén wave. For simplicity we take the limit of a *homogeneous plasma* and assume that $T_i \ll T_e$.

We start by generalizing the Boltzmann distribution (3.3) to the electromagnetic case by just adding $(e/m)(\mathrm{d}A_\parallel/\mathrm{d}t)$ to the right-hand side of equation (3.1). This leads to

$$\frac{\delta n_e}{n} = \frac{e\phi}{T_e} - \frac{\omega}{k_\parallel}\frac{eA_\parallel}{T_e}. \tag{3.47}$$

In the case of a homogeneous plasma we may completely neglect perpendicular electron motion. The electron continuity equation may then be written

$$\frac{\partial n_e}{\partial t} - \frac{1}{e}\nabla \cdot j_{\parallel e} = 0.$$

Now, neglecting parallel ion motion ($\omega \gg k_\parallel c_s$), we may write $j_{\parallel e} \approx j_\parallel$ and make use of equation (2.18). This leads to

$$\frac{\delta n_e}{n} = -\frac{k_\parallel k_\perp^2}{\omega}\frac{1}{\mu_0 e n}A_\parallel.$$

Since

$$\frac{1}{\mu_0 e n} = \frac{B^2}{\mu_0 n m_i}\frac{m_i}{eB^2} = \frac{v_A^2}{\Omega_{ci} B} = \rho^2 v_A^2 \frac{e}{T_e}$$

we obtain

$$\frac{\delta n_e}{n} = -k^2\rho^2\frac{k_\parallel}{\omega}v_A^2\frac{eA_\parallel}{T_e}. \tag{3.48}$$

Combining this result with equation (3.47) we obtain

$$A_\parallel = \frac{(k_\parallel/\omega)}{1 - k^2\rho^2(k_\parallel^2 v_A^2/\omega^2)}\phi.$$

Using this result instead of equation (2.19) in the derivation of equation (2.20) we obtain the dispersion relation of the kinetic Alfvén wave

$$\omega^2 = k_\parallel^2 v_A^2(1 + k^2\rho^2). \tag{3.49}$$

This is the simplest form of the kinetic Alfvén wave. The heating properties of this mode are related to singularities in the mode structure in plasmas with magnetic shear. The origin of such singularities is discussed in section 5.5.3. As is easily verified, the $k^2\rho^2$ part of equation (3.49) is due to a parallel electric field, which may be expressed as

$$E_\parallel = \mathrm{i}\frac{k^2\rho^2 k_\parallel^2 v_A^2}{\omega^2 - k^2\rho^2 k_\parallel^2 v_A^2} k_\parallel \phi.$$

A generalization of these results to inhomogeneous plasmas is made in exercise 8.

3.2.8 Quasi-linear diffusion

We shall now consider the particle transport due to low frequency modes in magnetized plasmas. We start by observing the correspondence between the continuity equation and the diffusion equation, i.e.

$$\frac{\partial n}{\partial t} + \nabla \cdot (n\boldsymbol{v}) = 0$$

may be written

$$\frac{\partial n}{\partial t} = -\nabla \cdot (\boldsymbol{\Gamma}) \tag{3.50}$$

where $\boldsymbol{\Gamma}$ is the flux, which according to Fick's law fulfils

$$\boldsymbol{\Gamma} = -D\nabla n \tag{3.51}$$

where D is the diffusion coefficient, and equation (3.50) reduces to the diffusion equation

$$\frac{\partial n}{\partial t} = \nabla \cdot (D\nabla n). \tag{3.52}$$

Equation (3.52) is only of interest in so far as it describes a secular steady state diffusion. We thus want to average equation (3.52) over the harmonic time and space variation of the fluctuations. In an inhomogeneous plasma a harmonic wave will always obtain a superimposed slow space variation of the amplitude due to the inhomogeneity, i.e.

$$\phi = \phi(x)\,\mathrm{e}^{-\mathrm{i}(\omega t - \boldsymbol{k}\cdot\boldsymbol{r})} + \mathrm{cc} \tag{3.53}$$

where the inhomogeneity is in the x direction. The flux in the x direction averaged over the harmonic variation is now

$$\langle\Gamma_x\rangle = \sum \delta n_k v_k^* + \mathrm{cc} \tag{3.54}$$

where

$$v_k = v_{Ex} = -\mathrm{i}\frac{k_y}{B}\phi_k.$$

The electrons are assumed to be close to Boltzmann distributed, but with a small imaginary correction due to dissipative effects, i.e.

$$\frac{\delta n_e}{n_0} = (1 - \mathrm{i}\delta)\frac{e\phi}{T_e}. \tag{3.55}$$

We may then write equation (3.54) as

$$\langle \Gamma_x \rangle = 2n_0 \frac{T_e}{eB} \sum_k \left| \frac{e\phi_k}{T_e} \right|^2 k_y \delta_k \tag{3.56}$$

or as a result of equation (3.51)

$$D_e = \frac{2}{\kappa} \frac{T_e}{eB} \sum \left| \frac{e\phi_k}{T_e} \right|^2 k_y \delta_k \tag{3.57}$$

where

$$\kappa = -\frac{1}{n_0} \frac{\mathrm{d}n_0}{\mathrm{d}x}.$$

We notice that the diffusion is due to the imaginary part of the deviation from the Boltzmann distribution. This dependence is such that unstable waves cause diffusion in positive x, i.e., towards the plasma boundary. As it turns out, δ_k will be proportional to κ in most cases of practical interest so that D remains finite when $\kappa \to 0$. It is also interesting to consider the ion diffusion. The ion density response is typically

$$\frac{\delta n_i}{n_0} = \frac{\omega_{*e}}{\omega_k} \frac{e\phi}{T_e} \tag{3.58}$$

in the region $\omega_k \gg k_\parallel c_s$, $k_\perp^2 c_s^2/\Omega_{ci}^2 \ll 1$. Equation (3.54) then becomes

$$\langle \Gamma_x \rangle = n_0 \frac{T_e}{eB} \sum_k \left| \frac{e\phi_k}{T_e} \right| \mathrm{i}k_y \frac{\omega_{*e}}{\omega_k} + \mathrm{CC}$$

with $\omega_k = \omega_r + \mathrm{i}\gamma_k$, and we now obtain

$$\langle \Gamma_x \rangle = 2n_0 \frac{T_e}{eB} \sum_k \left| \frac{e\phi_k}{T_e} \right|^2 \frac{k_y \gamma_k}{\omega_r^2 + \gamma_k^2} \omega_{*e}$$

and

$$D_i = 2\frac{T_e}{eB} \sum k_y \rho \frac{\gamma_k k_y c_s}{\omega_r^2 + \gamma_k^2} \left| \frac{e\phi}{T_e} \right|^2. \tag{3.59}$$

The instability saturates when κ has decreased to zero due to the diffusion. Another possible mechanism for saturation is when the nonlinear contribution to the radial derivative of the convective density perturbation becomes comparable to that of the linear, i.e., when $k_x \delta n = \kappa n_0$ or

$$\frac{\delta n}{n} = \frac{1}{k_x L_n} \tag{3.60}$$

where $L_n = 1/\kappa$ is the density inhomogeneity scale length. This is reasonable since the radial (x) inhomogeneity is driving the instability. This estimate is called the mixing length estimate.

Since the linear perturbation is

$$\delta n = -\boldsymbol{\xi} \cdot \nabla n_0 = -\xi_x \frac{\mathrm{d}n_0}{\mathrm{d}x} = \xi_x \frac{n_0}{L_n} \tag{3.61}$$

and

$$v_{kx} = -\mathrm{i}\omega\xi_x$$

we obtain from equation (3.54)

$$\Gamma_x = \frac{n_0}{L_n} \sum -\mathrm{i}\omega_k |\xi_x|^2 + \mathrm{cc.} \tag{3.62}$$

Now combining equations (3.60) and (3.61) we obtain

$$k_x |\xi_x| \leq 1. \tag{3.63}$$

This means that the displacement is a sizeable fraction of a wavelength. Now, using equation (3.63) in equation (3.62) we obtain the estimate

$$D = \Gamma \frac{L_n}{n_0} \leq 2 \sum \gamma_k / k_x^2.$$

If we interpret k_x as a correlation length of the full space variation we may omit the summation. The estimate is then actually written in the form

$$D \simeq \gamma / k_x^2 \tag{3.64}$$

This result can also be obtained by renormalization [7.3]. When the dominant nonlinearity is of the $E \times B$ convective type, as in the continuity or energy equation

$$\frac{\partial n}{\partial t} \sim v_{Ex} \frac{\partial}{\partial x} n$$

we can estimate the saturation level by balancing the linear growth, i.e., $\partial/\partial t \rightarrow \gamma$ with the nonlinearity. Then, representing ∇ by the inverse space scale of the full perturbation, the density (or temperature) perturbation cancels and we obtain the saturation level [3.44, 3.47]

$$\frac{e\phi}{T_e} \sim \frac{1}{k_x \rho} \frac{\gamma}{k_y c_s}. \tag{3.65}$$

Equation (3.63) is then replaced by

$$|\xi_x| \approx \frac{\gamma}{|\omega|}\frac{1}{k_x}. \tag{3.66}$$

Equation (3.66) shows that a real eigenfrequency reduces the step length because the convection oscillates in time. Using equation (3.66) in equation (3.62) we now obtain [3.44]:

$$D = \frac{\gamma^3/k_x^2}{\omega_r^2 + \gamma^2} \tag{3.67}$$

which turns into the mixing length estimate of equation (3.64) when $\gamma \gg \omega_r$. While equation (3.64) has the character of an upper limit, equation (3.67) is a more direct estimate of the transport level. Recently, a more general derivation has been made of equation (3.67) from a non-Markovian Fokker–Planck equation [3.48, 3.49]. It shows that equation (3.67) is a quite general expression, which only lacks off-diagonal elements. In the general expression, ω_r also contains the nonlinear frequency shift.

3.2.9 Confinement time

It is important to relate the diffusion coefficient to the confinement time τ. The confinement time is defined as the characteristic time for the decrease in total number of particles N, or total energy due to diffusion. For particle diffusion from a cylinder of radius r and length L we have

$$\tau = \frac{N}{(\mathrm{d}N/\mathrm{d}t)} = \frac{n\pi r^2 L}{2\pi r L \Gamma} = \frac{nr}{2\Gamma} \tag{3.68}$$

where n is the particle density and Γ is the particle flux given by equation (3.51). We then obtain

$$\tau = \frac{rL_n}{D} \tag{3.69}$$

where $L_n = -n/(\mathrm{d}n/\mathrm{d}t)$ is the characteristic length scale of density variation. If we take $L_n = r$ we find that for classical diffusion $\tau \sim r^2 B^2$, which would mean that by increasing the magnetic field we could build a smaller machine, obtaining the same confinement time. For quasi-linear diffusion, D will most probably scale as B^{-1} and a stronger increase in the magnetic field is necessary for a reduction in size. An increase in B also allows higher confined pressure and density. Clearly we can make the same derivation of the energy confinement time τ_E in terms of the thermal conductivity χ. As is evident from equation (1.8), the dependences of τ_E on a and R are much more complicated in a real toroidal system. The scaling with R seems to be, at least partly, due to the curvature radius of curvature-driven modes. It is, in fact, possible to obtain the scaling $R^{1.5}$ from equation (3.67) when the real eigenfrequency dominates and the growth rate is given by the root of κg.

3.3 Discussion

We have in the present chapter studied new eigenmodes associated with the inhomogeneity of a plasma. These modes are fundamental since they are the plasma's response to an inhomogeneity. They will, accordingly, have the effect of causing anomalous transport that tends to reduce the driving inhomogeneity, as also demonstrated. The modes studied were either of an MHD type with no parallel electric field, or of a drift type with electrostatic Boltzmann electrons. As seen from exercise 8, where a transition between these two types is made, the MHD-type modes are more global. They also generally have larger growth rates. In this chapter we have used a simple slab geometry to show the most fundamental properties of the modes. In chapter 5 more realistic geometries will be introduced. In chapter 4 we shall use a kinetic theory to re-derive dispersion relations for modes studied here.

3.4 Exercises

1. Explain why $\boldsymbol{v}_*$ does not contribute to equation (3.9).
2. (a) Generalize the derivation of equation (3.5) to the case of finite k_x. (b) Do the same with equation (3.11).
3. Discuss which of the effects included in equation (3.11) corresponds to compressibility.
4. An effect of finite Larmor radius (FLR) is that the ion particle $E \times B$ drift is reduced to

$$\boldsymbol{v}_{Ei} = \frac{1}{B}(\hat{\boldsymbol{z}} \times \nabla\phi)\left(1 - \frac{1}{2}k^2\rho_i^2\right)$$

As explained in section 2.3, we are allowed to replace fluid drifts by particle drifts throughout in the equation $\nabla \cdot \boldsymbol{j} = 0$. Use the FLR-corrected ion $E \times B$ drift to derive modified versions of the dispersion relations (3.27) and (3.28).
5. Show that by adding electron–ion collisions in the same way as in equation (3.17) we can obtain an instability driven by gravity in equation (3.40), i.e., for $\omega \ll k_\perp v_{\mathrm{th}\,e}$. This is a resistive interchange mode. The derivation may be simplified by assuming that $|k_y v_g| < |\omega_*|$ holds for both electrons and ions.
6. Assume a simple cylindrical geometry where the magnetic field is in the θ direction and the pressure decreases in the radial direction. Derive the β limit as a function of mode number m, $\phi \sim \mathrm{e}^{im\theta}$ when we make the replacement $\kappa \rightarrow (1/P)(\mathrm{d}P/\mathrm{d}r)$.
7. In a torus with both toroidal and poloidal magnetic field the magnetic field lines will move between the outside and inside of the torus. This effect is described by $q = \Delta\phi/\Delta\theta$, where $\Delta\phi$ and $\Delta\theta$ are the changes in toroidal and poloidal angle along a field line. We realize also that the relative direction between $\boldsymbol{g}$ and ∇n will change so that some regions have favourable, and some

unfavourable, curvature. In this case we may write

$$\frac{1}{R_c} = \frac{1}{R}\left(\cos\frac{2\pi z}{L} - \delta\right).$$

Since the space dependence along z is now no longer harmonic, we also have to make the replacement $k_\parallel \to -\mathrm{i}\partial/\partial z$. The Mathieu equation

$$\frac{\partial^2\phi}{\partial z^2} + \alpha\left(\cos\frac{2\pi z}{L} - \delta\right)\phi = 0$$

has the eigenvalue $\delta = \alpha L^2/8\pi$ for $\alpha L^2 \ll 2\pi^2$. Determine the β limit in the case $\beta L^2 \ll 2\pi^2 aR$, when $k_y^2 \gg k_x^2$, $\kappa \sim 1/a$, $\delta = a/2R$ and $L = 2\pi Rq$.

8. The approximation $E_\parallel = 0$ is one of the most frequently used approximations for flute modes ($k_\parallel \approx 0$). It is, however, not a good approximation for drift waves, which have a slightly larger $k_\parallel$. In order to see this it is necessary to consider the details of the electron dynamics. As is evident from the derivations of equations (2.20) and (3.28) we need only $A_\parallel$ in order to describe $\delta\boldsymbol{B}_\perp$.

(a) Show that the Boltzmann relation (3.3) is generalized to

$$\frac{\delta n_e}{n} = \frac{e\phi}{T_e} + \frac{\omega_{*e} - \omega}{k_\parallel}\frac{eA_\parallel}{T_e} \tag{3.70}$$

when we include $\delta\boldsymbol{B}_\perp = \nabla \times A_\parallel\hat{\boldsymbol{z}}$.

(b) Derive another expression for $\partial n_e/n$ from the electron continuity equation using

$$n_e v_{\parallel e} = -\frac{1}{e}j_{\parallel e} \approx -\frac{1}{e}j_\parallel$$

i.e., neglecting the parallel ion current, and express $j_\parallel$ in $A_\parallel$ by using Ampère's law. Put this expression equal to equation (3.70) and show that

$$E_\parallel = \mathrm{i}\frac{k^2\rho^2k_\parallel^2v_A^2}{\omega(\omega - \omega_{*e}) - k^2\rho^2k_\parallel^2v_A^2}k_\parallel\phi. \tag{3.71}$$

(c) Eliminate $A_\parallel$ and show that

$$\frac{\delta n_e}{n} = \left[1 + \frac{(\omega_{*e} - \omega)^2}{\omega(\omega_{*e} - \omega) + k^2\rho^2k_\parallel^2v_A^2}\right]\frac{e\phi}{T_e} \tag{3.72}$$

which turns into equation (3.25) for small $k_\parallel$, and into equation (3.3) for large $k_\parallel$.

9. Generalize equation (3.71) to include parallel ion motion when $v_{\parallel i}$ is given by $\partial v_{\parallel i}/\partial t = (e/m_i)E_\parallel$.

10. Use the tokamak parameters in appendix 1 to estimate for which mode number ($k_y \approx m/a$) the FLR correction to the β limit in equation (3.46) exceeds 20% of the β limit for low mode numbers.

11. Include the effects of finite ion Larmor radius in the dispersion relation (3.11).
12. Show that we recover the linear dispersion relation if we impose ambipolar electron and ion fluxes from equations (3.57) and (3.59).

References

[3.1] Suydam B R 1958 *Proc. UN Int. Conf. on Peaceful Uses of Atomic Energy* vol 31 (New York: Columbia University Press) p 157
[3.2] Mercier C 1960 *Nucl. Fusion* **1** 47
[3.3] Rudakov L I and Sagdeev R Z 1960 *Sov. Phys.–JETP* **37** 952
[3.4] Krall N A and Rosenbluth M N 1962 *Phys. Fluids* **5** 1435
[3.5] Mikhailovskii A B and Rudakov L I 1963 *Sov. Phys.–JETP* **17** 621
[3.6] Timofeev A V 1963 *Sov. Phys.–Tech. Phys.* **33** 776
[3.7] Kadomtsev B B 1965 *Plasma Turbulence* (New York: Academic)
[3.8] Krall N A 1966 *Phys. Fluids* **9** 820
[3.9] Moiseev S 1966 *JETP Lett.* **4** 55
[3.10] Kadomtsev B B and Pogutse O P 1967 *Sov. Phys.–JETP* **24** 1172
[3.11] Simon A 1968 *Phys. Fluids* **11** 1186
[3.12] Krall N A 1968 *Advances in Plasma Physics* vol 1, ed A Simon and W Tomphson (New York: Wiley) p 153
[3.13] Kadomtsev B B and Pogutse O P 1970 *Reviews of Plasma Physics* vol 5, ed M A Leontovich (New York: Consultants Bureau) p 249
[3.14] Hinton F L and Horton C W 1971 *Phys. Fluids* **14** 116
[3.15] Horton C W and Warma R K 1972 *Phys. Fluids* **15** 620
[3.16] Ichimaru S 1973 *Basic Principles of Plasma Physics: A Statistical Approach* (Reading: Benjamin) ch 8
[3.17] Krall N A and Trivelpiece A W 1973 *Principles of Plasma Physics* (New York: McGraw-Hill)
[3.18] Monticello D A and Simon A 1974 *Phys. Fluids* **17** 791
[3.19] Mikhailovskii A B 1974 *Theory of Plasma Instabilities* vol 2 (New York: Consultants Bureau)
[3.20] Connor J W and Hastie R J 1975 *Plasma Phys.* **17** 97, 109
[3.21] Guest G E, Hedrik C L and Nelson D B 1975 *Phys. Fluids* **18** 871
[3.22] Hasegawa A 1975 *Plasma Instabilities and Nonlinear Effects* (Berlin: Springer) ch 3
[3.23] Hasegawa A and Chen L 1975 *Phys. Rev. Lett.* **35** 370
[3.24] Berk H L 1976 *Phys. Fluids* **19** 1255
[3.25] Hasselberg G, Rogister A and El-Nadi A 1977 *Phys. Fluids* **20** 982
[3.26] Manheimer W M 1977 *An Introduction to Trapped Particle Instabilities in Tokamaks (ERDA Crit. Rev. Series)*
[3.27] Nichikawa K-I, Hatori T and Terashima Y 1978 *Phys. Fluids* **21** 1127
[3.28] Chen L, Hsu J and Kaw P K 1978 *Nucl. Fusion* **18** 1371
[3.29] Tang W M 1978 *Nucl. Fusion* **18** 1089
[3.30] Manheimer W M and Antonsen T M 1979 *Phys. Fluids* **22** 957
[3.31] Weiland J, Sanuki H and Liu C S 1981 *Phys. Fluids* **24** 98
[3.32] Weiland J 1981 *Phys. Scr.* **23** 801

[3.33] Rahman H U and Weiland J 1983 *Phys. Rev.* A **28** 1673
[3.34] Liewer P C 1985 *Nucl. Fusion* **25** 543
[3.35] Kaye S M 1985 *Phys. Fluids* **28** 2327
[3.36] Perkins F W and Sun Y C 1985 *Princeton Plasma Physics Laboratory Report* PPPL-2216
[3.37] Romanelli F, Tang W M and White R B 1986 *Nucl. Fusion* **26** 1515
[3.38] Rogister A, Hasselberg G, Kaleck A, Boileau A, van Andel H W H and von Hellerman M 1986 *Nucl. Fusion* **26** 797
[3.39] Waltz R E 1986 *Phys. Fluids* **29** 3684
[3.40] Tang W M 1986 *Nucl. Fusion* **26** 1605
[3.41] Tang W M, Rewoldt G and Chen L 1986 *Phys. Fluids* **29** 3715
[3.42] Dominquez R R and Waltz R E 1987 *Nucl. Fusion* **27** 65
[3.43] Jarmén A, Andersson P and Weiland J 1987 *Nucl. Fusion* **27** 941
[3.44] Weiland J and Nordman H 1988 *Proc. Varenna–Lausanne Workshop on Theory of Fusion Plasmas (Chexbres, 1988)* (Bologna: Editrice Compositori) p 451
[3.45] Weiland J 1988 *Comment. Plasma Phys. Control. Fusion* **12** 45
[3.46] Wootton A J *et al* 1988 *Plasma Phys. Control. Fusion* **30** 1479
[3.47] Horton W, Hong B G and Tang W M 1988 *Phys. Fluids* **31** 2971
[3.48] Zagorodny A, Weiland J and Jarmén A 1997 *Comment. Plasma Phys. Control. Fusion* **17** 353
[3.49] Zagorodny A and Weiland J 1998 *Ukr. J. Phys.* **43** 1402

Chapter 4

Kinetic Description of Low Frequency Modes in Inhomogeneous Plasma

4.1 Integration Along Unperturbed Orbits

In the previous chapter we derived simple dispersion relations for some of the most dangerous low frequency instabilities using a fluid description. We shall now show how this can be done using kinetic theory [4.1–4.18], from the Vlasov equation in a simple slab geometry. We shall start by using the method of integration along unperturbed orbits [4.1–4.5], which gives the most general result, i.e., also including modes with $\omega \geq \Omega_c$, full finite Larmor radius effects and wave–particle resonances. We shall, however, restrict our attention to modes with $\omega \ll \Omega_c$. We shall show how wave–particle resonances may impede free electron motion along the field lines, thus causing drift instability, and how the lowest order finite Larmor radius (FLR) effect agrees with that obtained from the stress tensor in chapter 2. After the more general treatment, we shall show how the wave–particle resonances can be described by a simpler drift-kinetic equation that does not contain FLR effects, and how the lowest order FLR effect can be obtained by a simple orbit averaging.

As indicated previously, the parallel motion of electrons may be affected by wave–particle resonances and also by inductance. We shall now give a kinetic description of these phenomena. The first problem that arises is to determine the unperturbed distribution function $f_0(v, x)$, where we choose the inhomogeneity to be in the x direction. For generality we also include in our description a gravitational force acting in the x direction. Inhomogeneities in the externally produced magnetic field may be included in this gravitational force, as well as a centrifugal force due to the toroidicity. The unperturbed distribution function $f_0(\boldsymbol{v}, x)$ may be written as a function of the constants of motion

$$W = \tfrac{1}{2}mv^2 - mgx \qquad P_y = m(v_y + \Omega_c x) \qquad \text{and} \qquad P_\parallel = mv_\parallel. \quad (4.1)$$

The constant P_y can easily be derived from the equation of motion in the form

$$\frac{\mathrm{d}\boldsymbol{v}}{\mathrm{d}t} = \Omega_c(\boldsymbol{v} \times \hat{\boldsymbol{z}})$$

or

$$\frac{\mathrm{d}v_y}{\mathrm{d}t} = -\Omega_c v_x = -\Omega_c \frac{\mathrm{d}x}{\mathrm{d}t}.$$

Here we assumed $\Omega_c = qB/m$ to be homogeneous. A *Maxwellian distribution function* corresponding to an exponential density gradient may be written

$$f_0(v, x) = n_0 \left(\frac{m}{2\pi T}\right)^{3/2} \exp\left[-\alpha'\left(x + \frac{v_y}{\Omega_c}\right)\right] \exp\left[-\frac{(1/2)mv^2 - mgx}{T}\right]. \tag{4.2}$$

It is important to note that Ω_c here contains the sign of the charge. For a plasma with perpendicular and parallel temperature gradients we may replace α' by $\alpha' + \delta_\perp v_\perp^2 + \delta_\parallel v_\parallel^2$. However, for simplicity we shall neglect temperature gradients. For a weak inhomogeneity, i.e., small α', we may expand the first exponential in equation (4.2). To first order in α' and g we then obtain the *zero order drift velocity*

$$\boldsymbol{v}_d = \frac{1}{n_0}\int \boldsymbol{v} f_0 \,\mathrm{d}\boldsymbol{v} = -\frac{\alpha' T}{m\Omega}\hat{\boldsymbol{y}} = -\left(\frac{\kappa T}{m\Omega_c} + \frac{g}{\Omega_c}\right)\hat{\boldsymbol{y}} \qquad \text{where } \kappa = -\frac{1}{n_0}\frac{\mathrm{d}n_0}{\mathrm{d}x}. \tag{4.3}$$

Note that in this expression we have to take Ω_c as negative for electrons. For simplicity we shall, in the following, continue to use the previous assumption that $k_x = 0$. We may then write

$$\boldsymbol{E}(\boldsymbol{r}, t) = \boldsymbol{E}_k \,\mathrm{e}^{\mathrm{i}(k_y y + k_\parallel z - \omega t)} + \mathrm{CC}$$
$$f(\boldsymbol{r}, \boldsymbol{v}, t) = f_k(\boldsymbol{v}) \,\mathrm{e}^{\mathrm{i}(k_y y + k_\parallel z - \omega t)} + \mathrm{CC}.$$

The linearized Vlasov equation may then be written

$$\left[\frac{\partial}{\partial t} + \boldsymbol{v}\cdot\frac{\partial}{\partial \boldsymbol{r}} + [\Omega_c(\boldsymbol{v}\times\hat{\boldsymbol{z}}) + g\hat{\boldsymbol{x}}]\cdot\frac{\partial}{\partial \boldsymbol{v}}\right] f_k(\boldsymbol{v})\,\mathrm{e}^{\mathrm{i}(k_y y + k_\parallel z - \omega t)}$$
$$= \frac{q}{m}(\boldsymbol{E}_k + \boldsymbol{v}\times\boldsymbol{B}_k)\cdot\frac{\partial f_0}{\partial \boldsymbol{v}}\,\mathrm{e}^{\mathrm{i}(k_y y + k_\parallel z - \omega t)}. \tag{4.4}$$

The left-hand side of equation (4.4) is here the total derivative along a particle orbit. In a linear approximation we use the unperturbed orbit

$$\boldsymbol{v}' = \boldsymbol{v}(t') = \tilde{B}(t' - t)[\boldsymbol{v}(t) - \boldsymbol{v}_g] + \boldsymbol{v}_g$$
$$\boldsymbol{r}' = \boldsymbol{r}(t) + \frac{1}{\Omega_c}\tilde{H}(t' - t)[\boldsymbol{v}(t) - \boldsymbol{v}_g] + \boldsymbol{v}_g(t' - t)$$

where $\boldsymbol{v}_g = -(g/\Omega_c)y$, and

$$\tilde{H}(t) = \begin{vmatrix} \sin\Omega_c t & 1 - \cos\Omega_c t & 0 \\ \cos\Omega_c t - 1 & \sin\Omega_c t & 0 \\ 0 & 0 & \Omega_c t \end{vmatrix}$$

and

$$\tilde{B} = \frac{1}{\Omega_c}\frac{\mathrm{d}\tilde{H}}{\mathrm{d}t}. \tag{4.5}$$

Integration along this orbit yields

$$f_k(\boldsymbol{v}) = \frac{q}{m}\int_0^\infty \mathrm{d}\tau\,(\boldsymbol{E}_k + \boldsymbol{v}\times\boldsymbol{B}_k)\cdot\frac{\partial f_0}{\partial \boldsymbol{v}}\,\mathrm{e}^{-\mathrm{i}\alpha(\tau)} \tag{4.6}$$

where

$$\alpha(\tau) = k_y\left[\frac{v_x}{\Omega_c}(1-\cos\Omega_c\tau) + \frac{v_y - v_g}{\Omega_c}\sin\Omega_c\tau + v_g\tau\right] + k_\parallel v_\parallel\tau - \omega\tau$$

and $\tau = t - t'$.

We shall here consider the region $\beta \ll 1$ ($\beta = 2\mu_0 nT/B^2$). In this region $k_\perp v_A \gg k_\perp v_{Di}$, and we may disregard the compressional (magnetosonic) wave. The only electromagnetic effect is then due to the bending of the magnetic field lines and we have

$$\delta\boldsymbol{B}_\parallel = (\nabla\times\boldsymbol{A})_\parallel = \frac{\partial A_x}{\partial y} - \frac{\partial A_y}{\partial x} = 0$$

for the perturbed field. This means that we can derive the perpendicular part of $\boldsymbol{A}$ from a potential, i.e., $\boldsymbol{A}_\perp = \nabla_\perp\chi$. Then from

$$\boldsymbol{E} = -\nabla\phi - \frac{\partial\boldsymbol{A}}{\partial t}$$

we find that we can write

$$\boldsymbol{E} = -\nabla_\perp\phi - \frac{\partial\psi}{\partial z}\hat{\boldsymbol{z}} \tag{4.7}$$

where

$$\phi = \phi + \frac{\partial\chi}{\partial t}$$

and

$$\psi = \phi + \int^z \frac{\partial A_\parallel}{\partial t}\,\mathrm{d}z.$$

From the Maxwell equation

$$\frac{\partial\boldsymbol{B}}{\partial t} = -\nabla\times\boldsymbol{E}$$

we then find

$$\frac{\partial\boldsymbol{B}}{\partial t} = (\hat{\boldsymbol{z}}\times\nabla_\perp)\frac{\partial}{\partial z}(\phi - \psi)$$

or for Fourier components

$$\boldsymbol{B}_k = \frac{\mathrm{i}}{\omega} k_y k_\| (\phi - \psi) \hat{\boldsymbol{x}}$$

where we introduced our assumption $k_x = 0$.

Observing that

$$\frac{\partial f_0}{\partial \boldsymbol{v}} = -\left(\frac{\alpha'}{\Omega_c}\hat{\boldsymbol{y}} + \frac{m}{T}\boldsymbol{v}\right) f_0(x, \boldsymbol{v}) \tag{4.8}$$

we may rewrite equation (4.6) as

$$\begin{aligned} f_k(\boldsymbol{v}) = & -\frac{q}{m}\int_0^\infty \mathrm{i}\left[\frac{m}{T}(k_y v_y' \phi + k_\| v_\|' \psi) + \frac{\alpha'}{\Omega_c} k_y \phi\right] f_0 \,\mathrm{e}^{-\mathrm{i}\alpha(\tau)}\,\mathrm{d}\tau \\ & + \mathrm{i}\frac{q}{m}\int_0^\infty \frac{\alpha'}{\Omega_c \omega} k_y k_\| (\boldsymbol{v} \times \hat{\boldsymbol{x}}) \cdot \hat{\boldsymbol{y}} (\phi - \psi) f_0 \,\mathrm{e}^{-\mathrm{i}\alpha(\tau)}\,\mathrm{d}\tau \end{aligned} \tag{4.9}$$

since

$$\frac{\mathrm{d}}{\mathrm{d}\tau}\mathrm{e}^{-\mathrm{i}\alpha(\tau)} = -\mathrm{i}[(k_y v_y' + k_\| v_\|') - \omega]\,\mathrm{e}^{-\mathrm{i}\alpha(\tau)}. \tag{4.10}$$

It is convenient to rewrite equation (4.9) as

$$\begin{aligned} f_k(\boldsymbol{v}) = & -\frac{q\phi}{T}\int_0^\infty [k_y v_y' + k_\| v_\|' - \omega] f_0 \,\mathrm{e}^{-\mathrm{i}\alpha(\tau)}\,\mathrm{d}\tau \\ & -\frac{q\phi}{T}\int_0^\infty \mathrm{i}(\omega - \omega_D) f_0 \,\mathrm{e}^{-\mathrm{i}\alpha(\tau)}\,\mathrm{d}\tau \\ & -\frac{q}{T}\int_0^\infty \mathrm{i}k_\| v_\|' (\psi - \phi) f_0 \,\mathrm{e}^{-\mathrm{i}\alpha(\tau)}\,\mathrm{d}\tau \\ & -\frac{q}{T}\int_0^\infty \mathrm{i}\frac{\omega_D}{\omega} k_\| v_\| (\phi - \psi) f_0 \,\mathrm{e}^{-\mathrm{i}\alpha(\tau)}\,\mathrm{d}\tau \end{aligned} \tag{4.11}$$

where

$$\omega_D = -k_y \frac{\alpha' T}{m\Omega} = \omega_* - k_y \frac{g}{\Omega_c}.$$

With the help of equation (4.10) we can immediately integrate the first integral of equation (4.11). We note that the limit $\tau = \infty$ corresponds to a contribution from the perturbation at $t' = -\infty$. We take this to be zero. Observing that the unperturbed distribution function is invariant along an unperturbed orbit, we then find

$$\begin{aligned} f_k(\boldsymbol{v}) = & -\frac{q}{T} f_0(x, \boldsymbol{v}) \left\{ \phi + \mathrm{i}\left[(\omega - \omega_D)\phi + k_\| v_\| (\psi - \phi)\left(1 - \frac{\omega_D}{\omega}\right)\right] \right. \\ & \left. \times \int_0^\infty \mathrm{e}^{-\mathrm{i}\alpha(\tau)}\,\mathrm{d}\tau \right\}. \end{aligned} \tag{4.12}$$

In order to evaluate the integral we now make use of the expansion

$$e^{-i\alpha(\tau)} = \sum_{n=-\infty}^{\infty} \sum_{n'=-\infty}^{\infty} J_n(\xi) J_n(\xi)' \exp\{i[n(\Omega_c\tau + \theta) - n'\theta + k_\parallel v_\parallel \tau - \tilde{\omega}\tau]\} \quad (4.13)$$

where $\tilde{\omega} = \omega + k_y g/\Omega_c$ and $\xi = k_y v_\perp/\Omega_c$.

We then obtain

$$f_k(\boldsymbol{v}) = -\frac{q}{T} f_0(x, \boldsymbol{v}) \left\{ \phi + \left[(\omega - \omega_D)\phi + k_\parallel v_\parallel (\psi - \phi) \left(1 - \frac{\omega_D}{\omega} \right) \right] \times \sum_{nn'} \frac{J_n(\xi) J_n(\xi)' e^{-i(n-n')\theta}}{n\Omega_c + k_\parallel v_\parallel - \tilde{\omega}} \right\}. \quad (4.14)$$

We may now obtain the dispersion relation from the Maxwell equations

$$\nabla \cdot \boldsymbol{E} = \frac{\rho}{\varepsilon_n} \quad (4.15)$$

$$(\nabla \times \boldsymbol{B})_\parallel = \mu_0 j_\parallel. \quad (4.16)$$

By using the formula

$$\int_0^\infty e^{-a^2x^2} x J_n(px) J_n(qx)\, dx = \frac{1}{2a^2} \exp\left[-\frac{p^2 + q^2}{4a^2} \right] I_n\left(\frac{pq}{2a^2} \right)$$

where I_n is a modified Bessel function, it is possible to show that

$$\sum_n \sum_{n'} \int \frac{f_0 J_n(\xi) J_n'(\xi) e^{-i(n-n')\theta}\, dv}{n\Omega + k_\parallel v_\parallel - \tilde{\omega}} = \sum_n \frac{\Lambda_n(\beta)}{\tilde{\omega} - n\Omega} \left[W\left(\frac{\tilde{\omega} - n\Omega}{|k_\parallel|(T/m)^{1/2}} \right) - 1 \right]$$

where $\Lambda_m(\beta) = I_n(\beta) e^{-\beta}$. $W(z)$ is the plasma dispersion function

$$W(z) = (2\pi)^{1/2} \int_{-\infty}^{\infty} \frac{x}{x - z} e^{-x^2/2}\, dx \qquad \text{and} \qquad \beta = k_\perp^2 T/m\Omega^2.$$

We thus obtain

$$n_k = \int f_k(\boldsymbol{v})\, dv = -\frac{qn_0}{T} \left\{ \phi + (\omega - \omega_D)\phi \times \sum_n \frac{\Lambda_n(\beta)}{\tilde{\omega} - n\Omega} \left[W\left(\frac{\tilde{\omega} - n\Omega}{|k_\parallel|(T/m)^{1/2}} \right) - 1 \right] + (\psi - \phi)\left(1 - \frac{\omega_D}{\omega} \right) \times \sum_n \Lambda_n(\beta) W\left(\frac{\tilde{\omega} - n\Omega}{|k_\parallel|(T/m)^{1/2}} \right) \right\}. \quad (4.17)$$

Similarly, we obtain

$$
\begin{aligned}
j_{\|k} &= q f_k(\boldsymbol{v}) v_z \, \mathrm{d}v \\
&= -\frac{q^2}{T}\frac{n_0}{k_\|}\left[(\omega-\omega_D)\phi \sum \Lambda_n(\beta) W\left(\frac{\tilde{\omega}-n\Omega}{|k_\||(T/m)^{1/2}}\right)\right. \\
&\quad \left. + (\psi-\phi)\left(1-\frac{\omega_D}{\omega}\right)\sum_n \Lambda_n(\beta)(\tilde{\omega}-n\Omega) W\left(\frac{\tilde{\omega}-n\Omega}{|k_\||(T/m)^{1/2}}\right)\right].
\end{aligned}
\tag{4.18}
$$

By using the density and current perturbations (4.17) and (4.18) we obtain the dispersion relation (4.16). In the following we shall mainly consider the frequency range $\tilde{\omega} \ll n\Omega$. In this region we need to include only the term $n = 0$ in the summations of equations (4.17) and (4.18). We further note that $\beta = k_\perp^2 T/m\Omega^2 = k_\perp^2\rho^2/2$ expresses the ratio between the Larmor radius ρ and the perpendicular wavelength. When β is small we may use the expansion

$$
\Lambda_0(\beta) = 1 - \beta \ (\beta \ll 1). \tag{4.19}
$$

The deviation of $\Lambda_0(\beta)$ from 1 will, in the following, be referred to as a finite Larmor radius effect. It is due to the fact that a particle that gyrates in a Larmor orbit on average experiences an electric field that is different from the field at the centre of the orbit. We shall always assume that $\Lambda_0(\beta) = 1$ for electrons, while we make different approximations for the ions.

We shall assume that condition (3.6) is also valid when we replace ω by the shifted frequency $\tilde{\omega} = \omega + k_y g/\Omega_c$. Using the expansions

$$
W(z) = \mathrm{i}\left(\frac{\pi}{2}\right)^{1/2} z\,\mathrm{e}^{-z^2/2} + 1 - z^2 + z^4/4 \ldots \qquad |z| \ll 1 \tag{4.20}
$$

and

$$
W(z) = \mathrm{i}\left(\frac{\pi}{2}\right)^{1/2} z\,\mathrm{e}^{-z^2/2} - \frac{1}{z^2} - 3/z^4 \ldots \qquad |z| \gg 1 \tag{4.21}
$$

and neglecting $k_y g/\Omega_{ce}$ for electrons (i.e., $\tilde{\omega} = \omega$, $\omega_{De} = \omega_{*e}$), we then obtain the particle densities

$$
\begin{aligned}
\frac{n_{ke}}{n_0} &= \frac{e}{T_e}\left\{\phi + \mathrm{i}\left(\frac{\pi}{2}\right)^{1/2} \frac{\omega-\omega_{*e}}{|k_\||(T_e/m_e)^{1/2}}\,\mathrm{e}^{-\omega^2/(k_\|^2 v_{te}^2)}\phi\right. \\
&\quad \left. + (\psi-\phi)\left(1-\frac{\omega_{*e}}{\omega}\right)\left[1+\mathrm{i}\left(\frac{\pi}{2}\right)^{1/2}\frac{\omega}{|k_\||(T_e/m_e)^{1/2}}\,\mathrm{e}^{-\omega^2/(k_\|^2 v_{te}^2)}\right]\right\}
\end{aligned}
\tag{4.22}
$$

$$\frac{n_{ki}}{n_0} = -\frac{e}{T_i}\left\{\phi - (\omega - \omega_{Di})\phi\frac{\Lambda_0(\beta_i)}{\tilde{\omega}}\right.$$
$$\times\left[1 + \frac{k_\parallel^2 T_i}{m_i\tilde{\omega}^2} - \mathrm{i}\left(\frac{\pi}{2}\right)^{1/2}\frac{\tilde{\omega}}{|k_\parallel|(T_e/m_e)^{1/2}}\,\mathrm{e}^{-\tilde{\omega}^2/(k_\parallel^2 v_{ti}^2)}\right]$$
$$+ (\psi - \phi)\left(1 - \frac{\omega_{Di}}{\omega}\right)\Lambda_0(\beta_i)$$
$$\left.\times\left[\mathrm{i}\left(\frac{\pi}{2}\right)^{1/2}\frac{\tilde{\omega}}{|k_\parallel|(T_e/m_e)^{1/2}}\,\mathrm{e}^{-\tilde{\omega}^2/(k_\parallel^2 v_{ti}^2)} - \frac{k_\parallel^2 T_i}{m_i\tilde{\omega}^2}\right]\right\} \tag{4.23}$$

and the parallel electron current

$$j_{\parallel ke} = -\frac{e^2 n_0}{T_e k_\parallel}\psi(\omega - \omega_{*e})\left[1 + \mathrm{i}\left(\frac{\pi}{2}\right)^{1/2}\frac{\omega}{|k_\parallel|(T_e/m_e)^{1/2}}\,\mathrm{e}^{-\omega^2/(k_\parallel^2 v_{te}^2)}\right]. \tag{4.24}$$

In equation (4.22) we recognize the Boltzmann distribution in the first term of the right-hand side. The second term represents the phase shift due to wave–particle resonance, and the third term is a correction due to induction.

4.1.1 Universal instability

The electrostatic limit is easily obtained from the above equations by setting $\psi = \phi$. The dispersion relation obtained from equation (4.15) can be written in the form

$$\varepsilon(k_y, k_\parallel, \omega) = 0 \tag{4.25}$$

where

$$\varepsilon = \frac{\phi_{\mathrm{ind}}}{\phi_{\mathrm{ext}}}.$$

Here ϕ_{ind} is the induced potential caused by an external potential ϕ_{ext}. In the region (3.6) we obtain

$$\varepsilon(k_y, k_\parallel, \omega) = 1 + \frac{k_{de}^2}{k_y^2}\left[1 + \mathrm{i}\left(\frac{\pi}{2}\right)^{1/2}\frac{\omega - \omega_{*e}}{|k_\parallel|(T_e/m_e)^{1/2}}\,\mathrm{e}^{-\omega^2/(k_\parallel^2 v_{te}^2)}\right]$$
$$+ \frac{k_{di}^2}{k_y^2}\left\{1 - \frac{\omega - \omega_{Di}}{\tilde{\omega}}\left[1 + \frac{k_\parallel^2 T_i/m}{\tilde{\omega}^2} - \mathrm{i}\left(\frac{\pi}{2}\right)^{1/2}\frac{\tilde{\omega}}{|k_\parallel|(T_e/m_e)^{1/2}}\right.\right.$$
$$\left.\left.\times\,\mathrm{e}^{-\tilde{\omega}^2/(k_\parallel^2 v_{ti}^2)}\right]\Lambda_0(\beta_i)\right\}. \tag{4.26}$$

In the limit $k\lambda_d \ll 1$, the 1 is negligible in equation (4.26) and the corresponding dispersion relation (4.25) can be obtained using quasi-neutrality. Assuming the wave–particle interaction to be weak we can treat imaginary parts of ε as perturbations. Thus, writing $\varepsilon = \varepsilon_R + \mathrm{i}\varepsilon_\perp$ we can solve equation (4.26) in the usual way as $\omega = \omega_r + \mathrm{i}\gamma$, where

$$\varepsilon_R(k_y, k_\parallel, \omega_r) = 0 \tag{4.27}$$

and

$$\gamma = -\frac{\varepsilon_I(k_y, k_\parallel, \omega_r)}{\partial\varepsilon_R/\partial\omega}. \tag{4.28}$$

From (4.27) we obtain the relation

$$1 + \frac{T_e}{T_i} - \frac{\omega - \omega_{*i}}{\tilde{\omega}}\left[\frac{T_e}{T_i} + \frac{k_\parallel^2 c_s^2}{\tilde{\omega}^2}\right]\Lambda_0(\beta_i) = 0. \tag{4.29}$$

For small $k_\parallel$ and dropping the gravity we obtain the solution

$$\omega = \frac{\omega_{*e}\Lambda_0(\beta_i)}{1 + (T_e/T_i)[1 - \Lambda_0(\beta_i)]} \tag{4.30}$$

where we use the relation

$$\omega_{*i} = -\frac{T_i}{T_e}\omega_{*e}.$$

By expanding $\Lambda_0(\beta_i)$ according to equation (4.19) and introducing $\beta_i = 1/2k_y^2\rho_i^2$, we obtain the solution

$$\omega = \omega_{*e}\left[1 - \frac{1}{2}k_y^2\rho_i^2\left(1 + \frac{T_e}{T_i}\right)\right]. \tag{4.31}$$

Since $\rho^2 = \frac{1}{2}(T_e/T_i)\rho_i^2$, we recognize the last term to be due to inertia (polarization drift) by comparison with equation (3.11). Clearly the ion inertia dominates the Larmor radius effect when $T_e \gg T_i$. From equation (4.28) we find

$$\gamma = -\left(\frac{\pi}{2}\right)^{1/2}\frac{\omega_r^2}{\omega_{*e}\Lambda_0(\beta_i)}\left[\frac{\omega - \omega_{*e}}{|k_\parallel|(T_e/m_e)^{1/2}}\,\mathrm{e}^{-\omega^2/(k_\parallel^2 v_{te}^2)} + \frac{T_e}{T_i}\frac{\omega_r - \omega_{*i}}{|k_\parallel|(T_i/m_i)^{1/2}}\,\mathrm{e}^{-\tilde{\omega}^2/(k_\parallel^2 v_{ti}^2)}\right]. \tag{4.32}$$

We notice that the situation concerning stability is similar to that for the influence of ion–electron collisions. As a result of Larmor radius effects we note from equation (4.31) that $\omega < \omega_{*e}$. According to equation (4.32), this means that the interaction between the wave and the electrons is destabilizing. Since $\omega_{*i} < 0$ for $\kappa > 0$, we find that the ions will cause damping. As a result of the exponential factor, however, this term will usually be small in the region (3.6). The collisionless instability described by equation (4.30) and equation (4.32) is usually referred to as a universal instability, since it was for a long time considered to be unavoidable in a finite size plasma. We note, however, that the Landau damping term will become important in a short device, where $k_\parallel$ has to take rather large values. Moreover, in a device with magnetic shear, $k_\parallel$ can take small values only locally and damping is obtained by convection into regions with larger $k_\parallel$. Finally, we note that since $\omega - \omega_{Di} = \tilde{\omega} - \omega_{*i}$, the only effect of gravity on the dispersion relation (4.29) is a shift of the real part of ω equivalent to a change of frame. Thus, in a frame moving with a velocity $\boldsymbol{v}_g = -g/\Omega_{ci}\hat{\boldsymbol{y}}$, the dispersion relation will take the same form as in the laboratory frame when the gravity is absent. This is, however, only true as long as we may drop the frequency-dependent terms in the electron part of the dispersion relation, i.e., as long as the electrons are able to maintain a Boltzmann distribution. For very small $k_\parallel$ this is no longer possible and we obtain a reactive instability known as a Rayleigh–Taylor or interchange instability.

4.1.2 Interchange instability

In the limit $k_\parallel \equiv 0$ we have $W(z_e) = W(z_i) = 0$. The electrostatic dispersion relation then takes the form

$$\frac{1}{T_e}\frac{\omega_{*e}}{\omega} = -\frac{1}{T_i}\left[1 - \frac{\tilde{\omega} - \omega_{*i}}{\tilde{\omega}}\Lambda_0\right]. \tag{4.33}$$

Introducing

$$\tilde{\omega} = \omega + k_y\frac{g}{\Omega_{ci}} = \omega - k_y v_{gi}$$

where $v_{gi} = -g/\Omega_{ci}$ is the gravitational drift and $\omega_{*i} = -(T_i/T_e)\omega_{*e}$, we may rewrite equation (4.33) as

$$\omega\omega_{*e}(1 - \Lambda_0) - \omega_{*e}k_y v_{gi} + \frac{T_e}{T_i}(1 - \Lambda_0)\omega(\omega - k_y v_{gi}) = 0. \tag{4.34}$$

Introducing

$$\omega_{*e} = k_y\frac{\kappa T_e}{eB_0}$$

in the constant term we finally arrive at the dispersion relation

$$\omega[\omega - k_y(v_{gi} + v_{*i})] = -\kappa g\frac{k_y^2\rho_i^2}{1 - \Lambda_0}. \tag{4.35}$$

This dispersion relation predicts instability when

$$\kappa g\frac{k_y^2\rho_i^2}{1 - \Lambda_0} > \frac{1}{4}k_y^2(v_{gi} + v_{*i})^2. \tag{4.36}$$

This instability is due to the charge separation created by a density perturbation when the electron and ion guiding centre drifts are different. Since $k_\parallel = 0$ the electrons cannot shield the charge separation by moving along B_0. This is the Rayleigh–Taylor or interchange instability. Clearly a condition to fulfil equation (4.35) is that $\kappa g > 0$, i.e., the gravity and density gradient are in opposite directions, as shown in figure 3.9.

In the unstable case the denser parts tend to change place (interchange), with the less dense parts thus causing convective diffusion. When ∇n and $\boldsymbol{g}$ have the same direction a perturbation is counteracted, which results in an oscillation. The instability is analogous to that of a heavy fluid resting on top of a light fluid. In a toroidal machine the centrifugal force due to the field line curvature may give rise to interchange instability in regions of unfavourable curvature ($\kappa g > 0$). We note also that when the different drift velocities of electrons and ions are caused by a gravitational force, the finite Larmor radius effect is stabilizing, contrary to the situation for drift waves.

4.1.3 Drift Alfvén waves and β limitation

We shall now consider the electromagnetic case in the region (3.6). We write the equations in a frame where the background guiding centre drift of the electrons is zero. In this frame the ion background drift will be equal to the difference between the ion and electron drifts in the laboratory frame. We shall assume that we can neglect the imaginary part of $W(z)$ for both ions and electrons. We then have from equations (4.20) and (4.21)

$$\frac{n_e}{n_0} = \frac{e}{T_e}\left[\phi + (\psi - \phi)\left(1 - \frac{\omega_{*e}}{\omega}\right)\right] \tag{4.37}$$

$$\frac{n_i}{n_0} = -\frac{e}{T_i}\left[\phi - (\omega - \omega_{Di})\frac{\Lambda_0}{\tilde{\omega}}\phi\right] \tag{4.38}$$

$$j_{\|_e} = \frac{e^2 n_0}{T_e k_\|}(\omega_{*e} - \omega)\psi. \tag{4.39}$$

By combining the induction law, equations (4.15) and (4.16), we obtain

$$j_\| = \frac{k_\| k_\perp^2}{\mu_0 \omega}(\phi - \psi). \tag{4.40}$$

Neglecting parallel ion motion we now obtain from equations (4.39) and (4.40)

$$\psi = \frac{k^2\rho^2 k_\|^2 v_A^2}{\omega(\omega_{*e} - \omega) + k^2\rho^2 k_\|^2 v_A^2}\phi. \tag{4.41}$$

Equation (4.41) gives the relation between the parallel and perpendicular electric fields. It involves the electron dynamics and can easily be obtained from the fluid equations. Now inserting equation (4.41) into equation (4.37) we obtain the electron response in terms of ϕ

$$\frac{n_e}{n_0} = \frac{e}{T_e}\phi\left[\frac{\omega_{*e}}{\omega} + \left(1 - \frac{\omega_{*e}}{\omega}\right)\frac{k^2\rho^2 k_\|^2 v_A^2}{\omega(\omega_{*e} - \omega) + k^2\rho^2 k_\|^2 v_A^2}\right]. \tag{4.42}$$

Equation (4.42), which is equivalent to equation (3.72), shows that for large $k_\parallel v_A$ the electron response approaches a Boltzmann distribution, while in the opposite case it is of the fluid mode type, i.e., proportional to ω_{*e}/ω. Introducing $\omega - \omega_{Di} = \tilde{\omega} - \omega_{*i}$ and $\omega_{*i} = -(T_i/T_e)\omega_{*e}$, we find from equation (4.38)

$$\frac{n_i}{n_0} = \frac{e}{T_e}\phi\left[\frac{\omega_{*e}}{\tilde{\omega}}\Lambda_0 - \frac{T_e}{T_i}(1-\Lambda_0)\right]. \tag{4.43}$$

We may now obtain a dispersion relation for drift Alfvén waves from equations (4.42) and equations (4.43) assuming quasi-neutrality. Multiplying by $\tilde{\omega}$ and using the definition $\tilde{\omega} = \omega - k_y v_g$ we obtain

$$\left(\omega_{*e} + \frac{T_e}{T_i}\tilde{\omega}\right)(\Lambda_0 - 1) + \omega_{*e}\frac{k_y v_g}{\omega} = (\omega - \omega_{*e})\frac{k^2\rho^2 k_\parallel^2 v_A^2}{\omega(\omega_{*e} - \omega) + k^2\rho^2 k_\parallel^2 v_A^2}.$$

We now multiply by the denominator on the right-hand side, divide by $(1 - \Lambda_0)$ and multiply by T_i/T_e. Observing now that $v_g = v_{gi} - v_{ge}$ and $v_{gj} = g_j/\Omega_{cj}$, we may write

$$\omega_{*e}k_y v_g = -k^2\rho^2\kappa g_i\left(1 + \frac{m_e g_e}{m_i g_i}\right) \tag{4.44}$$

and obtain the equation

$$\omega[\omega - k_y(v_g + v_{*i})] - k_\parallel^2 v_A^2 \frac{1}{2}\frac{k_\perp^2\rho_i^2}{1-\Lambda_0}\left(1 - \frac{k_y v_g}{\omega - \omega_{*e}}\right) + \kappa g_i\left(1 + \frac{m_e g_e}{m_i g_i}\right)\frac{1}{2}\frac{k_y^2\rho_i^2}{1-\Lambda_0}$$
$$= k_y^2\rho^2 k_\parallel^2 v_A^2 \frac{\omega - k_y(v_g + v_{*i})}{\omega - \omega_{*e}}. \tag{4.45}$$

Equation (4.45) is the dispersion relation for drift Alfvén waves including a gravitational force and full finite Larmor radius effects. We notice that equation (4.45) is identical to equation (4.35) in the limit $k_\parallel = 0$. We are, however, not allowed to take this limit of equation (4.45) since it would correspond to an expansion for $\omega/k_\parallel \ll v_{\mathrm{th}e}$. It gives the correct result because the electron density distributions become the same for the two cases, as seen from equations (4.42) and (4.17). We also notice that the flute mode response is obtained in the limit $\psi = 0$. In this case the induction force prevents the electrons from cancelling space charge by moving along B_0, and this makes the interchange mode solution possible. Clearly the Alfvén frequency $k_\parallel v_A$ has a stabilizing influence on the interchange instability. This can be seen as a result of the bending of the frozen in magnetic field lines, which counteracts the interchange of fluid elements. The balance between these forces leads to the β limit for stability discussed in chapter 2. The drift terms are also stabilizing. The $k_y v_{*i}$ term is due to a reduction of the convective $E \times B$ drift velocity of ions when averaged over a Larmor orbit, and leads to a stabilizing charge separation effect (compare section 3.2.6). The most unstable situation will obviously occur for small k_y.

If we expand equation (4.45) for small ion temperature, keeping only the lowest order Larmor radius effects and neglecting terms of order $k_y v_g/\omega$, we obtain the dispersion relation

$$\omega(\omega - \omega_{*i}) - k_\parallel^2 v_A^2 + \kappa g_i \left(1 + \frac{m_e g_e}{m_i g_i}\right) = k^2 \rho^2 k_\parallel^2 v_A^2 \frac{\omega - \omega_{*i}}{\omega - \omega_{*e}}. \tag{4.46}$$

Equation (4.46) agrees with equation (3.44) for small electron temperature if the gravity is expressed as a centrifugal acceleration, and thus verifies the lowest order finite Larmor radius effect as obtained from the stress tensor. The right-hand side of equation (4.46) is due to the parallel electric field and provides a coupling to the electrostatic drift wave branch. In studying this coupling we shall, for simplicity, neglect the gravity.

Assuming $k_\perp^2 \rho^2 \ll 1$ we then realize that equation (4.46) splits into two branches: the electric drift wave branch with $\omega = \omega_{*e}$, and the electromagnetic drift wave branch or drift Alfvén branch.

If we include the term proportional to $k_\parallel^2 T_i / m_i \omega^2$ in equation (4.23), equation (4.46) generalizes to

$$[\omega(\omega - \omega_{*e}) - k_\parallel^2 c_s^2][\omega(\omega - \omega_{*i}) - k_\parallel^2 v_A^2] = k_\perp^2 \rho^2 k_\parallel^2 v_A^2 (\omega - \omega_{*i})\omega. \tag{4.47}$$

This dispersion relation has four branches, as shown in figure 4.1. We notice from this figure, where $v_A > c_s$, that instead of an intersection of the branches corresponding to $k_\parallel > 0$ the branches change their identity and we obtain a region of strong coupling. The condition for the existence of this region is clearly $v_A > c_s$. On the other hand, in order to remain in the region (3.6) of drift waves we must have $v_A < v_{\text{th}\,e}$. This condition is equivalent to $\beta > m_e/m_i$, which is the limit of β where the electromagnetic effects have to be considered.

4.1.4 Landau damping

If we now return to equation (4.46), neglecting the right-hand side but including the electron and ion Landau damping effects from equations (4.22) and (4.23) to leading order, we obtain the dispersion relation

$$\omega(\omega - \omega_{*i}) - k_\parallel^2 v_A^2 + D = \mathrm{i}\omega\Gamma \tag{4.48}$$

where

$$D = \kappa g_i \left(1 + \frac{m_e}{m_i}\frac{g_e}{g_i}\right)$$

and

$$\Gamma = \sqrt{\pi/2} k_\parallel v_A \frac{v_A}{c_s}\left[\frac{\omega - \omega_{*i}}{\omega - \omega_{*e}}\left(\frac{T_e}{T_i}\right)^{3/2} \mathrm{e}^{-\tilde{\omega}^2/(k_\parallel^2 v_{ti}^2)} + \sqrt{m_e/m_i}\right].$$

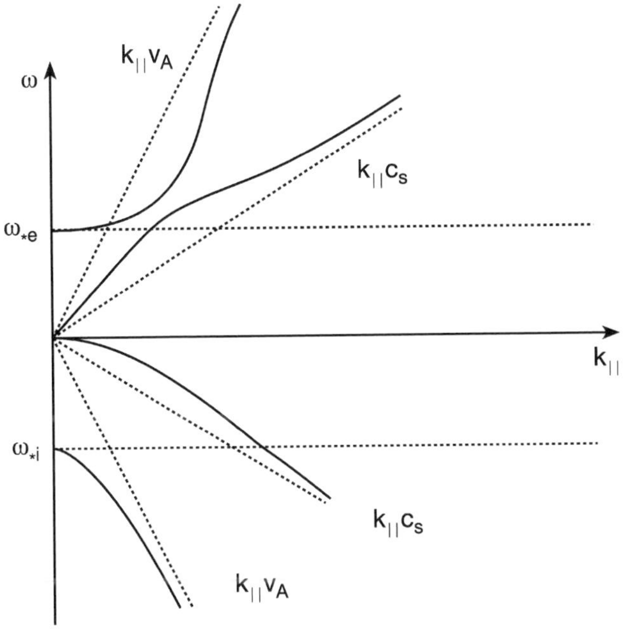

Figure 4.1. Dispersion diagram for electromagnetic drift waves.

Assuming now that $\omega = \omega_r + i\gamma$ and $\gamma \ll \omega_r$ we obtain by separation

$$\omega_r = \tfrac{1}{2}\omega_{*i} \pm \sqrt{\tfrac{1}{4}\omega_{*i}^2 + k_\parallel^2 v_A^2 - D} \tag{4.49}$$

$$\gamma = \frac{\omega_r \Gamma}{2\omega_r - \omega_{*i}}. \tag{4.50}$$

Here the sign of the denominator in equation (4.50) is given by the sign chosen for the root in equation (4.49). If

$$\tfrac{1}{4}\omega_{*i}^2 + k_\parallel^2 v_A^2 > D > k_\parallel^2 v_A^2$$

then the sign of ω_r does not change with the sign of the root and we can always find an unstable solution. This is exactly the region in which the MHD instability is stabilized by the FLR effect, so the *the dissipative effect restores the stability boundary to that of MHD*. We see, however, that the ion and electron contributions to Γ tend to cancel for $\omega_r \approx 1/2\omega_{*i}$, so the growth rate may be small in this region.

4.1.5 The magnetic drift mode

For the drift Alfvén wave we noticed that the electromagnetic effects disappear for $k_\parallel = 0$. There is, however, another mode that is electromagnetic and has $k_\parallel = 0$. This is the magnetostatic mode, which involves only electron motion. The electron motion along B_0 perturbs the magnetic field and the induction force acts back on the electrons. In a homogeneous plasma this mode is purely damped and has zero eigenfrequency. The perturbation of the magnetic field lines forms islands in the perpendicular plane and the motion of the electrons along the perturbed field lines causes anomalous heat transfer. In an inhomogeneous plasma, however, this mode has a frequency close to ω_{*e} and is no longer static. The mode is linearly described by a parallel induced electric field and a parallel vector potential A corresponding to a perpendicular magnetic field perturbation. We thus have

$$\boldsymbol{E} = E_\parallel \hat{\boldsymbol{z}} = -\frac{1}{c}\frac{\partial A}{\partial t}\hat{\boldsymbol{z}}. \tag{4.51}$$

Again assuming $k_x = 0$ we have

$$\boldsymbol{B} = B_x \hat{\boldsymbol{x}} = \mathrm{i}k_y A \hat{\boldsymbol{x}}. \tag{4.52}$$

Introducing these fields into (4.6) we obtain

$$\begin{aligned} f_k(\boldsymbol{v}) = {} & \frac{q}{m}\int_0^\infty \mathrm{i}\frac{m}{T}v'_\parallel \omega A f_0\,\mathrm{e}^{-\mathrm{i}\alpha(\tau)}\,\mathrm{d}\tau \\ & - \frac{q}{m}\int_0^\infty \mathrm{i}\frac{\kappa}{\Omega_c}k_y A(\boldsymbol{v}\times\hat{\boldsymbol{x}})\cdot\hat{\boldsymbol{y}} f_0\,\mathrm{e}^{-\mathrm{i}\alpha(\tau)}\,\mathrm{d}\tau \end{aligned} \tag{4.53}$$

which, observing that f_0 is invariant along the orbit, reduces to

$$f_k(\boldsymbol{v}) = \frac{q}{T} f_0 v_\parallel(\omega - \omega_*)A\int_0^\infty \mathrm{e}^{-\mathrm{i}\alpha(\tau)}\,\mathrm{d}\tau. \tag{4.54}$$

Making use of equation (4.13) we obtain

$$f_k(\boldsymbol{v}) = -\mathrm{i}\frac{q}{T}v_\parallel(\omega-\omega_*)A\sum_{nn'}\frac{J_n(\xi)J'_n(\xi)\,\mathrm{e}^{-\mathrm{i}(n-n')\theta}}{n\Omega_c - \omega}. \tag{4.55}$$

Assuming the average $v_\parallel$ to be zero, we now find the density perturbation to be zero. The parallel current is

$$j_{\parallel k} = q\int f_k v_\parallel\,\mathrm{d}v = -\frac{q^2}{T}n_0\frac{1}{2}v_{\mathrm{th}}^2 A(\omega-\omega_*)\sum_n\frac{\Lambda_n(\beta)}{n\Omega-\omega}. \tag{4.56}$$

Finally, we consider only the electron current, take only the term $n = 0$ and put $\Lambda_0(\beta) = 1$. Then using $T_e = (1/2)m_e v_{\mathrm{th}\,e}^2$ we find

$$j_{\parallel k} = -\frac{e^2 n_0}{m_e}\left(1 - \frac{\omega_{*e}}{\omega}\right)A. \tag{4.57}$$

Equation (4.16) for the present case takes the form

$$j_{\parallel k} = -\frac{1}{\mu_0}\frac{\partial^2 A}{\partial y^2}. \tag{4.58}$$

Combining equations (4.57) and (4.58) we have the dispersion relation

$$\omega = \frac{\omega_{*e}}{1 + k_y^2 c^2 / w_{pe}^2}. \tag{4.59}$$

4.2 The Drift Kinetic Equation

In the limit $k_\perp^2 \rho^2 \ll 1$ and $\omega \ll \Omega_c$, the previous procedure of integrating along the Larmor orbits can be avoided. The simplest possible approach in this limit is to write an equation of continuity for guiding centres. Such an equation can be written down immediately once the velocity and acceleration of the guiding centre is known. As it turns out, however, this method requires a more accurate knowledge of the guiding centre dynamics than more systematic procedures starting from the Vlasov equation, and it does not give an estimate of the magnitude of the neglected terms. In particular, it is difficult to obtain an accurate description of curvature effects. We shall thus restrict ourselves here to a slab geometry and leave the more complete description to a later systematic derivation.

The velocity of a guiding centre may be written

$$\boldsymbol{v}_{gc} = \frac{1}{B}(\boldsymbol{E} \times \boldsymbol{e}_\parallel) + v_\parallel \frac{\delta \boldsymbol{B}_\perp}{B} + \boldsymbol{v}_{gc}. \tag{4.60}$$

The acceleration is assumed only to be directed parallel to the magnetic field. The continuity equation may, as previously, be written in the form $\mathrm{d}f/\mathrm{d}t = 0$, which now becomes

$$\frac{\partial f}{\partial t} + (v_\parallel \boldsymbol{e}_\parallel + \boldsymbol{v}_{gc}) \cdot \nabla f + \frac{q}{m}[E_\parallel + (\boldsymbol{v}_{gc} \times \delta \boldsymbol{B}_\perp) \cdot \boldsymbol{e}_\parallel]\frac{\partial f}{\partial v_\parallel} = 0. \tag{4.61}$$

Since equation (4.61) no longer explicitly depends on $v_\perp$ we may integrate over the perpendicular velocity components. We thus have

$$f = f(\boldsymbol{r}, t, v_\parallel). \tag{4.62}$$

Equation (4.61) is the simplest form of the drift kinetic equation and does not contain finite Larmor radius effects. It does, however, keep the full parallel kinetic description and can be used to study wave–particle resonances. It is a simple exercise to rederive the dispersion relation (4.49) for the magnetic drift mode by using equation (4.61).

4.3 Dielectric Properties of Low Frequency Vortex Modes

We shall start by considering flute-like modes subject to the condition $k_\parallel^2 T_e/m\omega^2 < 1$. Dropping the Landau resonance terms but keeping electron FLR effects, we can write the dielectric function, observing that $\omega_P^2/\Omega_c^2 = \rho/\Lambda_0^2$ $(s_j = k^2 T_j/m_j \Omega_{cj}^2)$

$$\varepsilon(\omega, k_\parallel, k_\perp) = 1 + \frac{\omega_{pe}^2}{k_\perp^2 \rho_e^2 \Omega_{ce}^2}\left[1 - \frac{\omega - \omega_{*e}}{\omega}\left(1 + \frac{k_\parallel^2 T_e}{m_e \omega^2}\right)\Lambda_0(s_e)\right]$$
$$+ \frac{\omega_{pi}^2}{k_\perp^2 \rho_i^2 \Omega_{ci}^2}\left[1 - \frac{\omega - \omega_{*i}}{\omega}\left(1 + \frac{k_\parallel^2 T_i}{m_i \omega^2}\right)\Lambda_0(s_i)\right].$$

This expression has several interesting properties which we shall investigate. First, we expand for small Larmor radius and treat ω_*/ω, $k_\parallel T/m\omega^2$ and s as small. Then

$$\varepsilon(\omega, k_\parallel, k_\perp) = 1 + 2\frac{\omega_{pe}^2}{\Omega_{ce}^2}\left(\frac{1}{2} + \frac{\kappa}{k_\perp}\frac{\Omega_{ce}}{\omega} - \frac{k_\parallel^2}{k_\perp^2}\frac{\Omega_{ce}^2}{\omega^2}\right)$$
$$+ 2\frac{\omega_{pi}^2}{\Omega_{ci}^2}\left(\frac{1}{2} + \frac{\kappa}{k_\perp}\frac{\Omega_{ci}}{\omega} - \frac{k_\parallel^2}{k_\perp^2}\frac{\Omega_{ci}^2}{\omega^2}\right) \tag{4.63}$$

where $\kappa = -\mathrm{d}\ln n_0/\mathrm{d}x$.

For a tokamak plasma typically $\omega_{pe} \sim \Omega_{ce}$, while $\omega_{pi} \sim 50\Omega_{ci}$ (we observe also that $\omega_{pi}/\Omega_{ci} = c/v_A$). We notice that the commonly used expression

$$\varepsilon = \varepsilon_F = 1 + \frac{\omega_{pe}^2}{\Omega_{ce}^2} + \frac{\omega_{pi}^2}{\Omega_{ci}^2} \tag{4.64}$$

for flute modes in homogeneous plasmas is usually hard to fulfil in a realistic situation. In cases when the electron contribution can be dropped it may, however, sometimes be fulfilled. Assuming $k_\parallel = \kappa = 0$ and keeping the full FLR contribution we obtain

$$\varepsilon(k_\perp) = 1 + \frac{\omega_{pe}^2}{\Omega_{ce}^2}\frac{1 - \Lambda_0(s_e)}{s_e} + \frac{\omega_{pi}^2}{\Omega_{ci}^2}\frac{1 - \Lambda_0(s_i)}{s_i} \tag{4.65}$$

which shows that ε decreases for large Larmor radius.

The question of quasi-neutrality is also related to the dielectric constant ε_F. Assuming for instance

$$v_{\mathrm{th}\,i} \ll \frac{\omega}{k_\parallel} \ll v_{\mathrm{th}\,e}$$

and dropping the Landau resonances, parallel ion motion and Larmor radius, we obtain from equation (4.26)

$$\varepsilon(k_\perp, \omega) = 1 + \frac{1}{k_\perp^2 \Lambda_{de}^2}\left(1 + k_y^2\rho^2 - \frac{\omega_{*e}}{\omega}\right). \tag{4.66}$$

The dispersion relation for electrostatic drift waves, $\varepsilon(k_\perp, \omega) = 0$, can now be written

$$\omega = \frac{\omega_{*e}}{1 + k^2\rho^2[1 + (\lambda_{de}/\rho)^2]} = \frac{\omega_{*e}}{1 + k^2\rho^2(1 + \Omega_{ci}^2/\omega_{pi}^2)} \tag{4.67}$$

since $\lambda_{de}/\rho = \Omega_{ci}/\omega_{pi}$. The condition for applicability of quasi-neutrality is $k^2\lambda_d^2 \ll 1$, which leads us to drop the 1 in equation (4.66). This corresponds to dropping $\Omega_{ci}^2/\omega_{pi}^2$ in equation (4.67) which is equivalent to assuming that $\varepsilon_F \gg 1$. The condition for quasi-neutrality can be expressed as $\varepsilon_F \gg 1$ without involving the wavelength because we have compared the deviation from quasi-neutrality with the ion inertia term $k^2\rho^2$, which also contains the wavenumber.

The wave energy, as expressed by the formula

$$W = \frac{1}{4}\omega\frac{\partial\varepsilon}{\partial\omega}\langle E^2\rangle$$

is also closely related to the dielectric properties. We shall here consider the wave energy in two cases.

For the electrostatic drift waves, dropping gravity effects but keeping ion Larmor radius effects, we have

$$\varepsilon(\omega, k_\perp) = 1 + \frac{k_{de}}{k_\perp^2}\left[1 + \frac{T_e}{T_i}\left(1 - \frac{\omega - \omega_{*i}}{\omega}\Lambda_0(s)\right)\right]. \tag{4.68}$$

Assuming $k_\perp^2\lambda_{de} \ll 1$ (and $\varepsilon_F \gg 1$) we can write the dispersion relation

$$\omega = \omega_{*e}\Lambda_0(s)\left\{1 + \frac{T_e}{T_i}[1 - \Lambda_0(s)]\right\}^{-1}. \tag{4.69}$$

From equation (4.68) we also obtain

$$\frac{\partial\varepsilon}{\partial\omega} = \frac{k_{de}^2}{k_y^2}\frac{\omega_{*e}}{\omega^2}\Lambda_0(s). \tag{4.70}$$

Inserting equation (4.69) we then have

$$W_k = \frac{1}{4}k_{de}^2\left\{1 + \frac{T_e}{T_i}[1 - \Lambda_0(s)]\right\}|\phi_k|^2. \tag{4.71}$$

Here the second term is due to the ion polarization drift and tends to $k^2\rho^2$ in the limit $T_e/T_i \to \infty$.

For interchange modes ($k_\parallel = 0$) we write

$$\varepsilon(\omega, k_\perp) = 1 + \frac{k_{de}^2}{k_\perp^2}\left\{\frac{\omega_{*e}}{\omega} + \frac{T_e}{T_i}\left[1 - \frac{\tilde{\omega} - \omega_{*i}}{\tilde{\omega}}\Lambda_0(s)\right]\right\} \tag{4.72}$$

where $\tilde{\omega} = \omega - k_y v_g$.

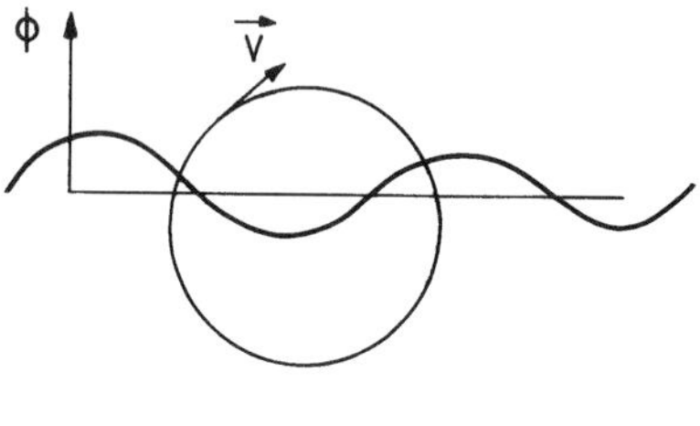

Figure 4.2. Finite gyroradius averaging.

Assuming quasi-neutrality we may write the dispersion relation

$$\frac{\omega_{*e}}{\omega} = -\frac{T_e}{T_i}\left[1 - \frac{\tilde{\omega} - \omega_{*i}}{\tilde{\omega}}\Lambda_0(s)\right]. \tag{4.73}$$

Multiplying with $\omega - k_y v_g$ we find

$$\frac{k_y v_g \omega_{*e}}{\omega} = \left(\frac{T_e}{T_i}\tilde{\omega} + \omega_{*e}\right)[1 - \Lambda_0(s)]. \tag{4.74}$$

Alternatively we may write equation (4.73) as

$$\frac{\omega_{*e}}{\tilde{\omega}}\Lambda_0(s) = \frac{\omega_{*e}}{\omega} - \frac{T_e}{T_i}[\Lambda_0(s) - 1]. \tag{4.75}$$

Differentiating ε we find

$$\frac{\partial\varepsilon}{\partial\omega} = \frac{k_{de}^2}{k_\perp^2}\left[\Lambda_0(s)\frac{\omega_{*e}}{\tilde{\omega}^2} - \frac{\omega_{*e}}{\omega^2}\right] \tag{4.76}$$

which, after substitution of equation (4.75), can be written

$$\frac{\partial\varepsilon}{\partial\omega} = \frac{k_{de}^2}{k_y^2}\frac{1}{\tilde{\omega}}\left[\frac{\omega_{*e}k_y v_g}{\omega^2} - \frac{T_e}{T_i}[\Lambda_0(s) - 1]\right]. \tag{4.77}$$

Then using equation (4.74) we find

$$\frac{\partial\varepsilon}{\partial\omega} = \frac{k_{de}^2}{k_\perp^2}\frac{1}{\tilde{\omega}}\left(\frac{T_e}{T_i}\frac{\tilde{\omega}}{\omega} + \frac{\omega_{*e}}{\omega} + \frac{T_e}{T_i}\right)[1 - \Lambda_0(s)] \tag{4.78}$$

from which

$$W = \frac{1}{4}k_{di}^2\frac{2\omega - \omega_{*i} - k_y v_g}{\omega - k_y v_g}[1 - \Lambda_0(s)]|\phi_k|^2. \tag{4.79}$$

Here we can see that the energy of the flute modes is an FLR effect.

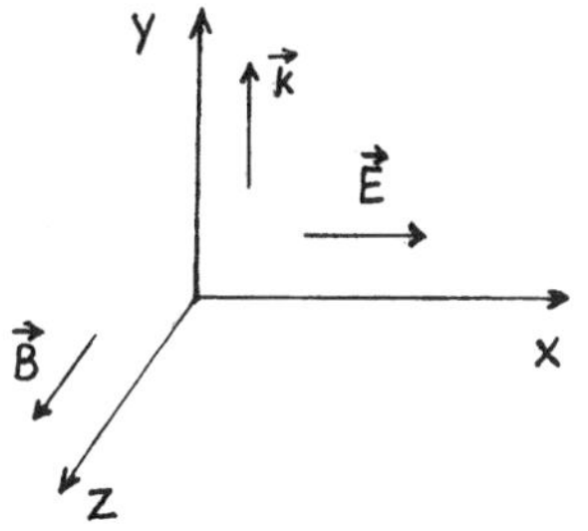

Figure 4.3. Slab geometry with electric field.

4.4 Finite Larmor Radius Effects Obtained by Orbit Averaging

In a fluid description the lowest order FLR effects can be obtained by including the diamagnetic and stress tensor drifts. Such a calculation, however, becomes rather involved due to cancellation between diamagnetic and stress tensor drifts that are not real particle drifts. Finite Larmor effects are due to the inhomogeneity of the electric field and the correction to the $E \times B$ drift caused by it. For a harmonic space dependence of the electric field and $\omega \ll \Omega_{ci}$, the FLR effect averages the electric field over a range of phases in space and this phase mixing always leads to reduction of the effective field (figure 4.2). The efficiency of this phase mixing clearly must depend on the ratio ρ/λ.

The particle equation of motion can be written

$$m\frac{\mathrm{d}\boldsymbol{v}}{\mathrm{d}t} = q[\boldsymbol{E} + \boldsymbol{v} \times \boldsymbol{B}]. \tag{4.80}$$

For simplicity we use a slab geometry according to figure 4.3, where $\boldsymbol{B} = B_0\hat{\boldsymbol{z}}$ and

$$\boldsymbol{E} = E_0 \cos(ky - \omega t)\hat{\boldsymbol{x}}. \tag{4.81}$$

In component form we have

$$\frac{\mathrm{d}v_x}{\mathrm{d}t} = \Omega_c v_y + \frac{q}{m}E_0 \cos(ky(t) - \omega t) \tag{4.82}$$

$$\frac{\mathrm{d}v_y}{\mathrm{d}t} = -\Omega_c v_x \tag{4.83}$$

where we observe that the electric field is evaluated at $y(t)$, i.e., along the orbit. The coupling between the equations on the time scale ω_c^{-1} can be eliminated by differencing with respect to t and substitution. This leads to

$$\frac{\mathrm{d}^2 v_x}{\mathrm{d}t^2} = -\Omega_c^2 v_x + \frac{q}{m}\left(\omega - k\frac{\mathrm{d}y}{\mathrm{d}t}\right) E_0 \sin[ky(t) - \omega(t)] \tag{4.84}$$

$$\frac{d^2 v_y}{dt^2} = -\Omega_c^2 v_y - \Omega_c^2 \frac{1}{B_0} E_0 \cos[ky(t) - \omega(t)]. \tag{4.85}$$

We shall now assume that $\Omega_c \gg \omega$ so that the time scales are well separated. We then average over the short time scale, obtaining

$$\langle v_x \rangle = \frac{1}{B_0 \Omega_c} E_0 \left\langle \left(\omega - k\frac{dy}{dt} \right) \sin[ky(t) - \omega(t)] \right\rangle - \frac{1}{\Omega_c^2} \left\langle \frac{d^2 v_x}{dx^2} \right\rangle \tag{4.86}$$

$$\langle v_y \rangle = -\frac{1}{B_0} E_0 \langle \cos[ky(t) - \omega(t)] \rangle - \frac{1}{\Omega_c^2} \left\langle \frac{d^2 v_y}{dx^2} \right\rangle. \tag{4.87}$$

We shall now perform the averaging of equations (4.86) and (4.87) over the unperturbed orbit, obtained by solving equation (4.68) with $E_0 = 0$. This orbit may be written

$$y(t) = y_0 + r_L(t) \tag{4.88}$$

where

$$r_L(t) = -\frac{v_\perp}{\Omega_c} [\cos(\Omega_c t + \phi) - \cos\phi] \tag{4.89}$$

is the projection of the Larmor radius along y. The orbit in equation (4.89) corresponds to

$$v_y = v_\perp \sin(\Omega_c t + \phi)$$
$$v_x = v_\perp \cos(\Omega_c t + \phi).$$

For the orbit (4.89) we have $\langle d^2 v_{x,y}/dt^2 \rangle = 0$. We are also interested in the lowest order FLR effects and take only linear terms in the parameter $k^2 r_L^2 \ll 1$. We then have

$$\begin{aligned} \sin[ky(t) - \omega t] = {} & \sin[ky_0 - \omega t] \left\{ 1 - \frac{1}{2} \frac{k^2 v_\perp^2}{\Omega_c^2} [\cos(\Omega_c t + \phi) - \cos\phi]^2 \right\} \\ & - \cos(ky_0 - \omega t) \frac{k v_\perp}{\Omega_c} [\cos(\Omega_c t + \phi) - \cos\phi)] \\ \cos(ky(t) - \omega t) = {} & \cos(ky_0 - \omega t) \left\{ 1 - \frac{1}{2} \frac{k^2 v_\perp^2}{\Omega_c^2} [\cos(\Omega_c t + \phi) - \cos\phi]^2 \right\} \\ & + \sin[ky_0 - \omega t] \frac{k v_\perp}{\Omega_c} [\cos(\Omega_c t + \phi) - \cos\phi)]. \end{aligned}$$

We now perform the averaging over time, keeping ωt constant since $\Omega_c \gg \omega$. We then obtain

$$\begin{aligned} \langle \sin[ky(t) - \omega t] \rangle = {} & \sin[ky_0 - \omega t] \left[1 - \frac{1}{2} \frac{k_\perp^2 v_\perp^2}{\Omega_c^2} \left(\frac{1}{2} + \cos^2\phi \right) \right] \\ & + \cos(ky_0 - \omega t) \frac{k v_\perp}{\Omega_c} \cos\phi \end{aligned} \tag{4.90}$$

$$\langle\cos(ky(t)-\omega t)\rangle = \cos(ky_0-\omega t)\left[1-\frac{1}{2}\frac{k_\perp^2 v_\perp^2}{\Omega_c^2}\left(\frac{1}{2}+\cos^2\phi\right)\right]$$
$$-\sin[ky_0-\omega t]\frac{kv_\perp}{\Omega_c}\cos\phi. \tag{4.91}$$

We also need

$$\left\langle\frac{\mathrm{d}y}{\mathrm{d}t}\sin[ky(t)-\omega t]\right\rangle = 0.$$

We now continue to average equation (4.91) over a Maxwellian distribution. Then $v_\perp^2 = 2T/m \cos\phi = 0$, $\cos^2\phi = 1/2$ and

$$\overline{\langle\sin[ky(t)-\omega t]\rangle} = \sin(ky_0-\omega t)\left(1-k^2\frac{T}{m\Omega_c^2}\right)$$
$$\overline{\langle\cos[ky(t)-\omega t]\rangle} = \cos(ky_0-\omega t)\left(1-k^2\frac{T}{m\Omega_c^2}\right).$$

Introducing $\rho^2 = 2T/m\Omega_c^2$ we now have

$$\langle\overline{v_x}\rangle = \frac{\omega}{B_0\Omega_c}E_0\sin[ky_0-\omega t]\left(1-\frac{1}{2}k^2\rho^2\right) \tag{4.92}$$
$$\langle\overline{v_y}\rangle = \frac{1}{B_0}E_0\cos(ky_0-\omega t)\left(1-\frac{1}{2}k^2\rho^2\right) \tag{4.93}$$

which are the averaged drifts we have been seeking. With the present choice of $\boldsymbol{E}$ and $\boldsymbol{k}$ we have the $E \times B$ drift in the y direction and the polarization drift in the x direction. As it turns out, the present averaging is not accurate enough to give correct FLR correction to the polarization drift. Thus, if the perturbed orbit is introduced into $\langle \mathrm{d}^2 v_{x,y}/\mathrm{d}t^2\rangle$, we obtain new terms of the same order as the FLR correction to equation (4.92).

Neglecting this effect we obtain in vector form

$$\langle\boldsymbol{v}_\perp\rangle = (1-\tfrac{1}{2}k^2\rho^2)\boldsymbol{v}_E + \boldsymbol{v}_p. \tag{4.94}$$

We observe here that the constant of integration, $\cos\phi$, in equation (4.89) is important in order to include all particles with orbits through y_0. Since $y = y_0 + v_\perp/\Omega_c \cos\phi$, the representation (4.89) means that we include particles with gyrocentres between $y_0 - v_\perp/\Omega_c$ and $y_0 + v_\perp/\Omega_c$.

In order to compare our results with those from a fluid theory we now calculate a density response to an electric field by using the continuity equation. This is a natural procedure since the density response is uniquely defined, while fluid and particle drifts may differ. Thus using equation (4.94) for ions in the continuity equation, neglecting parallel ion motion, we obtain for $k \gg |\nabla \ln n_0|$

$$\frac{\delta n}{n} = \left[\frac{\omega_{*e}}{\omega}\left(1-\frac{k^2T_i}{m_i\Omega_{ci}^2}\right)-\frac{k^2T_e}{m_i\Omega_{ci}^2}\right]\frac{e\phi}{T_e} \tag{4.95}$$

where the last term is due to the polarization drift. Equation (4.95) can also be rewritten in the form

$$\frac{\delta n_i}{n} = \left[\frac{\omega_{*e}}{\omega} - \frac{k^2 T_e}{m_i \Omega_{ci}^2} \left(1 - \frac{\omega_{*i}}{\omega} \right) \right] \frac{e\phi}{T_e} \tag{4.96}$$

where the FLR term now appears as the $-\omega_{*i}/\omega$ correction to the polarization drift. Equation (4.96) can be obtained by using fluid equations and including the diamagnetic drift in the convective derivative in the polarization drift, i.e.

$$\frac{\partial}{\partial t} \rightarrow \frac{\partial}{\partial t} + \boldsymbol{v}_{*j} \cdot \nabla.$$

This is also what remains in the fluid description after cancellations between diamagnetic and stress tensor drifts (compare chapter 2). It can also be readily shown by the orbit averaging method that this procedure can also be used for the perturbed diamagnetic drift, thus giving the lowest order nonlinear FLR effects when used in the convective derivative in the polarization drift.

It is also interesting to note the similarity between the FLR effects and the polarization drift in their contribution to the density response. Such a similarity may be expected since the FLR effect is due to the space dependence of the electric field along the orbit, while the polarization drift is due to the time dependence. A particle gyrating in the orbit cannot distinguish between these origins of field variation.

4.5 Discussion

We have, in this chapter, rederived the dispersion relations of chapter 3 using a kinetic description. This has been simplified by using a slab geometry. A more general gyrokinetic description will be given in chapter 5. We have also considered particularly the effects of finite Larmor radius and verified the first order effects that were obtained from fluid theory in chapter 2, and the consequences of it for stability found in chapter 3. Finally, the dielectric properties of inhomogeneous plasmas are fundamental. Later, in chapter 7, we shall show how the wave energy of interchange modes can be recovered to first order in the FLR parameter from a nonlinear conservation relation. In chapter 5 we shall use more realistic geometries and also study modes driven by temperature gradients.

4.6 Exercises

1. Perform the integration in equation (4.3).
2. Derive a relation between ϕ and ψ in equation (4.7) using fluid equations and neglecting parallel ion motion. Compare with the expression for $E_\parallel$ in equation (3.71).

3. Derive equation (4.47) by using fluid equations.
4. Derive a dispersion relation for the universal drift instability for small Larmor radius by using equation (4.22) for electrons and fluid equations for ions, i.e., neglecting ion Landau damping.
5. Use the tokamak data in appendix 1 to compare the different contributions to ε_F in equation (4.64).

References

[4.1] Rosenbluth M N and Longmire C 1957 *Ann. Phys., NY* **1** 120

[4.2] Rosenbluth M N and Rostoker N 1959 *Phys. Fluids* **2** 23

[4.3] Rudakov L I and Sagdeev R Z 1960 *Sov. Phys.–JETP* **37** 952

[4.4] Rosenbluth M N, Krall N A and Rostoker N 1962 *Nucl. Fusion* (Suppl.) **1** 143

[4.5] Krall N A and Rosenbluth M N 1962 *Phys. Fluids* **5** 1435

[4.6] Kadomtsev B B 1965 *Plasma Turbulence* (New York: Academic)

[4.7] Sitenko A G 1967 *Electromagnetic Fluctuations in a Plasma* (New York: Academic)

[4.8] Krall N A 1968 *Advances in Plasma Physics* vol 1, ed A Simon and W Thompson (New York: Wiley) p 153

[4.9] Coppi B, Laval G, Pellat R and Rosenbluth M N 1968 *Plasma Phys.* **10** 1

[4.10] Rukhadze L I and Silin V P 1969 *Sov. Phys.–Usp.* **2** 659

[4.11] Kadomtsev B B and Pogutse O P 1970 *Reviews of Plasma Physics* vol 5, ed M A Leontovitch (New York: Consultants Bureau) p 249

[4.12] Ichimaru S 1973 *Basic Principles of Plasma Physics* (New York: McGraw-Hill)

[4.13] Krall N A and Trivelpiece A W 1973 *Principles of Plasma Physics* (New York: McGraw-Hill)

[4.14] Mikhailovskii A B 1974 *Theory of Plasma Instabilities* vol 2 (New York: Consultants Bureau)

[4.15] Hasagawa A 1975 *Plasma Instabilities and Nonlinear Effects* (Berlin: Springer) ch 3

[4.16] Manheimer W M 1977 *An Introduction to Trapped Particle Instabilities in a Tokamak (ERDA Crit. Rev. Series)* (Energy Research and Development Administration)

[4.17] Gary S P and Sanderson J J 1978 *Phys. Fluids* **21** 1181

[4.18] Hasegawa A 1979 *Phys. Fluids* **22** 1988

Chapter 5

Low Frequency Modes in Inhomogeneous Magnetic Fields

We have seen how some typical low frequency modes can be driven unstable by density, pressure or current gradients in simple geometries. A more accurate description of collective modes in magnetic confinement systems requires, in general, more detailed geometry effects, as well as separate effects of density and temperature gradients [5.1–5.186]. In the present chapter we shall aim to make the geometrical description more accurate, thus, in most cases, leading to eigenvalue problems for the modes concerned. We shall also derive a more complete drift kinetic description, introduce the gyrokinetic equation and present an advanced fluid model. We shall furthermore briefly review the fields of transport due to magnetic fluctuations and advanced fluid models.

5.1 Anomalous Transport in Systems with Inhomogeneous Magnetic Fields

Although work on understanding transport in magnetic confinement systems has been going on for about 50 years, this problem is still a major scientific issue [5.167]. Its importance for the size and cost of a reactor is obvious and critical, but the scientific difficulties associated with it are enormous.

Initially, for ohmically heated plasmas, the interest was mainly focused on electron transport since it was dominant. While particle transport has to be ambipolar, energy transport does not. Thus, electron thermal transport through magnetic perturbations is an obvious option. The best known scaling law in this regime is the Alcator scaling [5.31]:

$$\tau_e \sim na^2. \tag{5.1}$$

Several theories have been able to recover the density dependence through a dependence on collisionality. A candidate for magnetic transport is the microtearing mode [5.57, 5.58], while the dissipative trapped electron mode [5.10] can give such a scaling through electrostatic dynamics [5.85]. We shall return later to trapped electron modes and briefly discuss magnetic perturbations here. Several papers consider transport for given magnetic fluctuations (or islands). The island width, which determines whether islands from neighbouring rational surfaces overlap, is critical for transport. When this is the case we can use the Rechester–Rosenbluth diffusion coefficient [5.41]:

$$D = v_{\mathrm{th}\,e} L_c \left(\frac{\delta B}{B} \right)^2 \tag{5.2}$$

where L_c is the correlation length, which, in general, depends on the resistivity through the mode width. An obvious candidate for creating magnetic perturbations is the magnetic drift mode of equation (4.59). This mode has $k_\parallel = 0$ and $\phi = 0$. In a realistic plasma with magnetic shear, this mode is localized near rational surfaces (see the following section), where $k_\parallel \approx 0$. In such a geometry we have a radial eigenvalue problem with the boundary conditions $\phi = 0$ (odd ϕ) and even $A_\parallel$ at the rational surface. Such a mode is a tearing mode [5.4], which is destabilized by resistivity.

While most confinement systems are designed to eliminate the dangerous global tearing modes with $k_{\mathrm{pol}} L_n \leq 0$, a localized microtearing mode with $k_{\mathrm{pol}} L_n \gg 1$ can still be unstable if $\nu_{ei} > \omega_{*e}$. The saturation level due to diffusion is [5.57]:

$$\frac{\delta B}{B} = \frac{\rho_e}{L_{T_e}}. \tag{5.3}$$

This level is of the order of 10^{-5} to 10^{-4} in typical tokamaks. This mode is thus a candidate for explaining the Alcator scaling in the ohmic regime, which is usually collision dominated. It is, however, almost stable in collisionless plasmas, giving a very small transport.

In the collisionless regime, electromagnetic drift wave turbulence has been considered as a candidate for generating magnetic transport [5.34, 5.88, 5.97]. The magnetic fluctuation level is, however, usually too low, or the correlation length too short due to very high mode numbers. A remaining possibility is nonlinearly self-sustained magnetic perturbations. As an example, collisionless tearing modes can be driven unstable by the turbulent radial diffusion of electrons [5.119]. The experimental situation remains unclear. On the one hand, evaluations of the magnetic flutter transport on TEXT [5.125] conclude that it is considerably smaller than the total transport, while experiments on Tore Supra [5.126] indicate the presence of magnetic islands. A recent development in this field is the current diffusive ballooning mode [5.124]. It is an MHD type mode which is described by resistive MHD equations. A transport model has been based on this mode. It tends to give good agreement with experiments for electron thermal diffusion but not such good agreement for ion diffusion [5.182]. This may be due to the fact that only one fluid equation is used and a full kinetic derivation is still lacking.

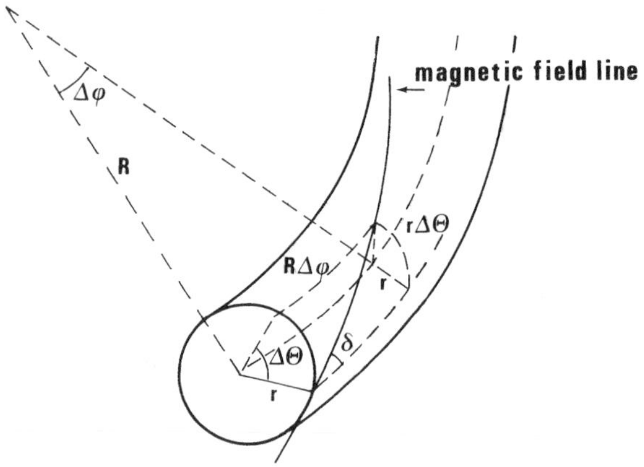

Figure 5.1. Toroidal geometry.

When the density is increased sufficiently, the confinement time saturates and another instability takes over. Transport code simulations [5.86, 5.87] indicate that this is the ion temperature gradient driven mode [5.1, 5.28, 5.61].

5.2 Toroidal Mode Structure

A general plasma perturbation in a torus must, in order to fulfil the boundary conditions, be a superposition of elementary perturbations of the form

$$f(r, \theta, \phi) = \overline{f}(r)\, \mathrm{e}^{\mathrm{i}(m\theta - n\phi)} \tag{5.4}$$

where θ is the poloidal and ϕ is the toroidal angle according to figure 5.1.

Here the phase angle can be represented as

$$m\theta - n\phi = k_\theta r\theta + k_\phi R\phi$$

where

$$k_\theta = \frac{m}{r} \qquad \text{and} \qquad k_\phi = -\frac{n}{R}.$$

The magnetic field can be written

$$\boldsymbol{B} = B_\theta \boldsymbol{e}_\theta + B_\phi \boldsymbol{e}_\phi$$

where

$$\boldsymbol{k} \cdot \boldsymbol{B} = \frac{m}{r} B_\theta - \frac{n}{R} B_\phi.$$

We now introduce

$$q(r) = \frac{\Delta\phi}{\Delta\theta} = \frac{B_\phi}{B_\theta}\frac{r}{R} \tag{5.5}$$

where $\Delta\phi$ and $\Delta\theta$ are changes in ϕ and θ on a translation along a field line, as shown in figure 5.1. We can then write

$$\boldsymbol{k} \cdot \boldsymbol{B} = \frac{n}{r}B_\theta\left[\frac{m}{n} - q(r)\right] \tag{5.6}$$

showing that $\boldsymbol{k} \cdot \boldsymbol{B} = 0$ when $q(r) = m/n$. This means that the pitch angle of an equi-phase line $\alpha = (r/R)m/n$ coincides with the pitch angle of the magnetic field lines $\delta = r\Delta\theta/R\delta\phi = r/qR$. In this situation $k_\parallel = 0$, and the electrons cannot cancel the space charge caused by the mode (m, n) on the magnetic surface (the surface containing magnetic field lines) corresponding to $q(r) = m/n$. This surface is called the rational surface. Since $q(r)$ usually grows monotonically with r, each mode will not have more than one rational surface. Modes that are well localized around the rational surface are usually more unstable since the effective $k_\parallel$ is small. One common way of expressing $k_\parallel$ is by rewriting equation (5.6) as

$$k_\parallel = \frac{\boldsymbol{k} \cdot \boldsymbol{B}}{B} = \frac{n}{r}\frac{B_\theta}{B_\phi}\left[\frac{m}{n} - q(r)\right] = [m - nq(r)]/Rq \tag{5.7}$$

where we assume that $B_\theta \ll B_\phi$ so that $B \approx B_\phi$. The wave number $1/Rq$ represents the inhomogeneity of the magnetic field and is related to the connection length L_c, defined by

$$L_c = 2\pi Rq.$$

L_c is a measure of the length of a magnetic field line between two points with the same θ. We shall introduce

$$k_c = \frac{2\pi}{L_c} = \frac{1}{qR}$$

so that

$$k_\parallel = [m - nq(r)]k_c. \tag{5.8}$$

Taylor expanding $q(r)$ around the rational surface and introducing $q(r_0) = m/n$ we obtain

$$k_\parallel = -nk_c\frac{\mathrm{d}q}{\mathrm{d}r}(r - r_0) = -\frac{r - r_0}{L_s}k_\theta \tag{5.9}$$

where we introduced the shear length

$$L_s = \frac{k_\theta}{k_c}\frac{1}{r(\mathrm{d}q/\mathrm{d}r)}. \tag{5.10}$$

A frequently used measure of the shear strength is also

$$s = \frac{\mathrm{d}\ln q}{\mathrm{d}\ln r} = \frac{r}{q}\frac{\mathrm{d}q}{\mathrm{d}r}. \tag{5.11}$$

These two parameters are related through

$$L_s = \frac{Rq}{s} = (sk_c)^{-1}.$$

For a tokamak, typically, s is small near the axis and is otherwise of order 1. Another quantity which is often of interest is the distance between neighbouring rational surfaces. If $q(r_0) = m/n$ and $q(r_0 + \Delta r) = (m+1)/n$ we obtain for slowly varying $q(r)$

$$\Delta r = \left(n\frac{\mathrm{d}q}{\mathrm{d}r}\right)^{-1}.$$

If we vary n instead, an additional factor q will appear. Since a mode is usually localized around its rational surface, the question of overlapping between two modes and, accordingly, nonlinear interaction and transport properties, depends strongly on the distance between rational surfaces.

Another inhomogeneity which is important for the mode structure is the decrease of B_ϕ along the main radius. This variation can be expressed as

$$B_\phi = \frac{B_T}{1 + (r/R)\cos\theta}. \tag{5.12}$$

Modes which are driven by the curvature of the magnetic field lines are usually strongly influenced by the different sign of the curvature on the outside and inside of the torus, introducing a periodicity with period L_c along the magnetic field lines. Although it is still possible to Fourier decompose the modes into components of the type in equation (5.4), these components will be linearly coupled and an eigenmode will now have the form

$$f(r,\theta,\phi) = \bar{f}(r,\theta)\,\mathrm{e}^{\mathrm{i}(m\theta - n\phi)}. \tag{5.13}$$

This leads to a two-dimensional problem for the mode structure, which in general is difficult to treat exactly, and approximate analytical solutions are usually only available if the r or θ dependence dominates.

The poloidal variation of f is often, by projection, transferred to a variation along the magnetic field. A Fourier decomposition along the magnetic field then leads to a coupling between components with different $k_\parallel$. Since a convection in the radial direction changes $k_\parallel$, we realize that we shall obtain a coupling between the mode structure along the magnetic field and the position in the radial direction. This coupling usually tends to inhibit the radial convection, thus reducing the shear damping. Since $f(r,\theta)$ will vary at least as fast as the fundamental mode $m = 1$, it is not necessary to distinguish between the modes

m and $m+1$. It is instead common to express an eigenmode by its independent mode number n. Close to the rational surface the poloidal variation is described by $m = q(r)n$ and an additional variation of f. Due to the poloidal variation we must, however, also introduce a θ-dependent safety factor $\nu(r, \theta)$ so that

$$q(r) = \frac{1}{2\pi} \oint \nu(r, \theta)\, \mathrm{d}\theta.$$

The representation (5.13) then turns into

$$f(r, \theta, \phi) = \bar{f}(r, \theta) \exp\left\{ in \left[\int^{\theta} \nu(r, \theta)\, \mathrm{d}\theta - \phi \right] \right\}. \tag{5.14}$$

This is a very useful eikonal description, which, for large n, describes a mode with a rapid variation across the magnetic field and a slow variation along the magnetic field. The r-dependent helicity that results from putting $m = q(r)n$ corresponds to a mode that tends to follow the field lines when it moves in the radial direction, thereby minimizing $k_{\parallel}$ and the restoring line bending force. There is, however, one disadvantage, which has been discussed extensively in connection with electromagnetic ballooning modes. It is the lack of periodicity of the phase function at a distance from the rational surface. This cannot be compensated by the amplitude f within the eikonal description. The problem was solved by transforming the problem to an infinite domain in θ where no periodicity is required, and then constructing a periodic solution by adding the integer Fourier components [5.39, 5.42].

5.2.1 Curvature relations

We shall now discuss some fundamental relations obtained in a curved magnetic field from a fluid point of view. We start with the condition for pressure balance. By adding the equations of motion for ions and electrons and dropping the inertial terms (error of order ω/Ω_c) we obtain

$$\nabla p = \boldsymbol{j} \times \boldsymbol{B}. \tag{5.15}$$

Combining equation (5.15) with

$$\nabla \times \boldsymbol{B} = \mu_0 \boldsymbol{j} \tag{5.16}$$

we obtain the pressure balance equation

$$\nabla \left(p + \frac{1}{2\mu_0} B^2 \right) = \frac{1}{\mu_0} (\boldsymbol{B} \cdot \nabla) \boldsymbol{B} \tag{5.17}$$

where $(1/2\mu_0)B^2$ is the magnetic field pressure and $(\boldsymbol{B} \cdot \nabla)\boldsymbol{B}$ is the field curvature. When written for the background quantities, equation (5.17) shows how the magnetic field pressure varies in space due to particle pressure (diamagnetic effect) and field curvature. In a low β plasma the ∇p term is often neglected and equation (5.17) then just gives the geometrical relation of the vacuum field. If we, on the other hand, write equation (5.17) for perturbed quantities and linearize we observe that

$$(\boldsymbol{B} \cdot \nabla)\boldsymbol{B} \sim \boldsymbol{B}_0 k_\parallel \delta \boldsymbol{B} + (\delta \boldsymbol{B} \cdot \nabla)\boldsymbol{B}_0 \sim \frac{1}{R}\delta B B_0$$

for $k_\parallel \sim 1/r$, where R is the radius of curvature of the background field. This estimate is typical for quasi-flute modes in toroidal machines and since $\nabla B^2 \sim k_\perp B_0 \delta B$ we realize that the curvature term is normally negligible for perturbations. We then have

$$\nabla\left(\delta p + \frac{1}{2\mu_0}\delta B^2\right) = 0.$$

Since $\delta B^2 = \delta(\boldsymbol{B}_0 + \delta \boldsymbol{B})^2 \approx 2\boldsymbol{B}_0 \cdot \delta \boldsymbol{B}$ we find the relation

$$\delta B_\parallel = -\frac{\mu_0 \delta p}{B_0} \tag{5.18}$$

which relates the parallel perturbation in B to the pressure perturbation. We now return to the derivation of the drift velocities in chapter 2. Introducing $\boldsymbol{e}_\parallel = \boldsymbol{B}_0/B_0$ we have

$$\boldsymbol{e}_\parallel \times (\boldsymbol{v} \times \boldsymbol{B}) = \boldsymbol{v} B_\parallel - \boldsymbol{B} v_\parallel = \boldsymbol{v}_\perp B_\parallel = \boldsymbol{v}_\perp B_0 \left(1 + \frac{\delta B_\parallel}{B_0}\right).$$

Then linearizing the expression for $\boldsymbol{v}_\perp$ we find, dropping $\boldsymbol{v}_\pi$ and $\boldsymbol{v}_g$, that the only drift which is modified is $\boldsymbol{v}_*$. The quantity usually needed in the derivation of dispersion relations is $\nabla \cdot \boldsymbol{j}_\perp$. We are interested in evaluating the expression

$$\begin{aligned}
&\nabla \cdot \left[(n_i \boldsymbol{v}_{*i} - n_e \boldsymbol{v}_{*e})\left(1 - \frac{\delta B_\parallel}{B_0}\right)\right] \approx \nabla \cdot \left[\frac{1}{eB_0}(\boldsymbol{e}_\parallel \times \nabla p)\left(1 - \frac{\delta B_\parallel}{B_0}\right)\right] \\
&\approx -\frac{1}{eB_0}\frac{\nabla B_0}{B_0} \cdot (\boldsymbol{e}_\parallel \times \nabla \delta p) + \frac{1}{eB_0}\nabla \cdot (\boldsymbol{e}_\parallel \times \nabla \delta p) - \frac{1}{eB_0}(\boldsymbol{e}_\parallel \times \nabla p_0) \cdot \nabla \frac{\delta B_\parallel}{B_0} \\
&= \frac{1}{eB_0}\left(\boldsymbol{e}_\parallel \times \frac{\delta B_\parallel}{B_0}\right) \cdot \nabla \delta p + \frac{1}{eB_0}(\nabla \times \boldsymbol{e}_\parallel) \cdot \nabla \delta p + \frac{2\mu_0}{eB_0^3}(\boldsymbol{e}_\parallel \times \nabla p_0) \cdot \nabla \delta p
\end{aligned} \tag{5.19}$$

where we started by assuming $\delta B_\parallel \ll B_0$, then used quasi-neutrality, linearized and finally used equation (5.17), assuming that

$$\frac{\nabla \delta B_\parallel}{\delta B_\parallel} \gg \frac{\nabla B_0}{B_0}.$$

We shall now rewrite $\nabla \times \boldsymbol{e}_\parallel$ using the vector relations of appendix 2. From equation (8.6) we have

$$\nabla(\boldsymbol{e}_\parallel \cdot \boldsymbol{e}_\parallel) = 2\boldsymbol{e}_\parallel \times (\nabla \times \boldsymbol{e}_\parallel) + 2(\boldsymbol{e}_\parallel \cdot \nabla)\boldsymbol{e}_\parallel = 0.$$

Taking the vector product with $\boldsymbol{e}_\parallel$ we find

$$(\nabla \times \boldsymbol{e}_\parallel)_\perp = \boldsymbol{e}_\parallel \times (\boldsymbol{e}_\parallel \cdot \nabla)\boldsymbol{e}_\parallel. \tag{5.20}$$

Since

$$\boldsymbol{e}_\parallel \cdot (\nabla \times \boldsymbol{e}_\parallel) = \frac{1}{B}\boldsymbol{e}_\parallel \cdot (\nabla \times \boldsymbol{B})$$

is associated with a background current, and since $k_\parallel$ is generally assumed to be small we shall neglect the parallel component of $\nabla \times \boldsymbol{e}_\parallel$. We can now use equation (5.17) for background fields to express the first term in equation (5.19) in the two others. It then turns out that the finite beta terms cancel. Then introducing the curvature vector

$$\boldsymbol{\kappa} = (\boldsymbol{e}_\parallel \cdot \nabla)\boldsymbol{e}_\parallel = -\frac{\boldsymbol{R}_c}{R_c^2}$$

we can write

$$\nabla \cdot \left[(n_i \boldsymbol{v}_{*i} - n_e \boldsymbol{v}_{*e})\left(1 - \frac{\delta B_\parallel}{B_0}\right)\right] \approx \frac{2}{eB_0}(\boldsymbol{e}_\parallel \times \boldsymbol{\kappa}) \cdot \nabla \delta p. \tag{5.21}$$

This result has several interesting implications. First, as we already noticed, the finite β terms cancel, making a low β treatment adequate. Second, we see that the divergence of the diamagnetic drift flux is a curvature effect (compare equation (2.8)). The term given by equation (5.21) is in fact the leading order curvature effect in an expansion in a/R (inverse aspect ratio), and is the main driving pressure term for ballooning modes. It can be represented by an equivalent gravity drift and this gives the same result as that obtained for the kinetic derivation of interchange modes. The drift terms $k_y v_g$ in the derivation are, however, higher order in a/R and do not correctly describe the effect of a curved field. The result (5.21) suggests that we introduce an effective curvature drift

$$\boldsymbol{v}_{\kappa j} = 2\frac{T_j}{q_j B_0}(\boldsymbol{e}_\parallel \times \boldsymbol{\kappa}) \tag{5.22}$$

which is the effective total magnetic drift in a Maxwellian plasma, including the lowest order finite β effects.

When effects of $\delta B_\parallel$ are not included we have the curvature relations

$$\nabla \cdot (n\boldsymbol{v}_*) = \frac{1}{T}\boldsymbol{v}_D \cdot \nabla \delta P \tag{5.23}$$

and

$$\nabla \cdot \boldsymbol{v}_E = \frac{q}{T}\boldsymbol{v}_D \cdot \nabla \phi \tag{5.24}$$

where

$$\boldsymbol{v}_D = \frac{T}{m\Omega_C}\left(\boldsymbol{e}_\parallel \times \frac{\nabla B}{B}\right) + \frac{T}{m\Omega_c}(\boldsymbol{e}_\parallel \times \boldsymbol{\kappa}) \tag{5.25}$$

is the sum of the ∇B drift and the curvature drift. These drifts are the same as the kinetic ∇B and the curvature drifts when those are averaged over a Maxwellian distribution, i.e., $\langle v_\parallel^2 \rangle = T/m$, $\langle v_\perp^2 \rangle = 2T/m$. From the comparison we see that if the fluid equations were generalized to a situation with different $T_\parallel$ and $T_\perp$, we should use $T_\perp$ for the ∇B drift and $T_\parallel$ for the curvature drift.

For anisotropic temperature we, in fact, get a contribution from the curvature drift to the fluid drift. It is [5.137]

$$\boldsymbol{v}_{D\,\text{fluid}} = \frac{T_\parallel - T_\perp}{T_\parallel}\boldsymbol{v}_\kappa \tag{5.26}$$

where $\boldsymbol{v}_\kappa = (T_\parallel/m\Omega_c)\boldsymbol{e}_\parallel \times \boldsymbol{\kappa}$. Moreover, the diamagnetic heat flow is split into two parts [5.137]:

$$\boldsymbol{q}_*^\parallel = \frac{1}{2}\frac{P_\perp}{m\Omega_c}\boldsymbol{e}_\parallel \times \nabla T_\parallel + (P_\parallel - P_\perp)\boldsymbol{v}_\kappa \tag{5.27}$$

$$\boldsymbol{q}_*^\perp = 2\frac{P_\perp}{m\Omega_c}\boldsymbol{e}_\parallel \times \nabla T_\perp. \tag{5.28}$$

For isotropic pressure these add up to the Braginskii $\boldsymbol{q}_*$.

5.3 The Influence of Magnetic Shear on Drift Waves

As pointed out in the previous section, in a tokamak the magnetic field has both a toroidal and a poloidal component. Moreover, since the poloidal field is generated by the toroidal plasma current it varies with r. Assuming, for instance, a homogeneous current density and applying Ampère's law to a circular contour with radius r around the centre of the plasma in the perpendicular plane, we find $B_p = \mu_0(1/2)jr$, where j is the current density. It is thus natural to assume that B_p increases with r. In our previous Cartesian coordinate system, the x coordinate corresponds to r and the y coordinate to the poloidal direction. The simplest possible approximation of the magnetic field is now

$$\boldsymbol{B}(x) = B_0\left(\hat{z} + \frac{x}{L_s}\hat{y}\right) \tag{5.29}$$

where L_s is the characteristic scale length of the magnetic field variation. It usually fulfils $L_s/a \gg 1$, where a is the small radius. This kind of transverse variation of the magnetic field is referred to as magnetic shear. In order to describe drift waves in a system with magnetic shear we have to solve a differential equation for the field variation in x, and the solution for the mode frequency becomes an eigenvalue problem. We now consider perturbations of the form

$$f(x, y, z, t) = \tilde{f}(x)\,\mathrm{e}^{\mathrm{i}(k_y y + k_\parallel z - \omega t)} \tag{5.30}$$

where f may represent any perturbed quantity. We may then write the perpendicular ion velocity, including v_E and v_{pi}, from equations (2.3) and (2.4) as

$$\boldsymbol{v}_{\perp i} = \frac{1}{B_0}\frac{\partial \phi}{\partial x}\hat{\boldsymbol{y}} - \frac{\mathrm{i}k_y}{B_0}\phi\hat{\boldsymbol{x}} + \frac{\omega}{B_0\Omega_i}\left(\mathrm{i}\frac{\partial \phi}{\partial x}\hat{\boldsymbol{x}} - k_y\phi\hat{\boldsymbol{y}}\right). \tag{5.31}$$

The ion continuity equation now yields

$$\frac{\delta n_i}{n_0} = \left(\frac{\omega_{*e}}{\omega} + \frac{T_e}{eB_0\Omega_{ci}}\frac{\partial^2}{\partial x^2} - \frac{k_y^2 T_e}{eB_0\Omega_{ci}} + \frac{T_e}{m_i}\frac{k_\parallel^2}{\omega^2}\right)\frac{e\phi}{T_e}. \tag{5.32}$$

We shall, for simplicity, disregard destabilizing effects and use the approximation

$$\frac{\delta n_e}{n_0} = \frac{e\phi}{T_e} \tag{5.33}$$

for the electron density. We now want to introduce the leading order effect of the magnetic shear into the system described by equations (5.31) and (5.32). The effect of the magnetic shear will be to twist the magnetic field. A toroidal eigenmode will also be twisted according to its poloidal and toroidal mode numbers. At a certain value of r it has the same degree of twisting as the magnetic field and $k_\parallel = 0$. At larger r the poloidal field will have a projection on z. The simplest model for its variation in a Cartesian system is (compare equation (5.6)).

$$k_\parallel = \frac{x}{L}k_y. \tag{5.34}$$

Introducing equation (5.33) into equation (5.31) and using the quasi-neutrality condition we obtain the eigenvalue equation

$$\rho^2\frac{\partial^2 \phi}{\partial x^2} + \frac{c_s^2}{v_{*e}^2}\frac{x^2}{L_s^2}\phi + \left(\frac{\omega_{*e}}{\omega} - 1 - k_y^2\rho^2\right)\phi = 0 \tag{5.35}$$

where we approximated ω by $k_y v_{*e}$ in the term proportional to x^2, since this term is assumed to be small. Equation (5.35) has a solution of the form

$$\phi = H_n(\mathrm{i}\xi)\,\mathrm{e}^{\pm \mathrm{i}\xi^2/2} \tag{5.36}$$

where H_n is a Hermite polynomial of order n and

$$\xi = \left(\frac{\Omega_{ci}}{v_{*e}L_s}\right)^{1/2} x.$$

If equation (5.36) is substituted into equation (5.35) we obtain the condition

$$\frac{v_{*e}\Omega_{ci}L_s}{c_s^2}\left(\frac{\omega_{*e}}{\omega}-1-k_y^2\rho^2\right)=\pm(2n+1) \tag{5.37}$$

which determines the eigenvalue ω. Clearly, the $\pm$ in equations (5.36) and (5.37) is related to the direction of propagation of the wave. Assuming the presence of absorbing boundaries the group velocity must be outward. Since this corresponds to an inward phase velocity we have to choose the minus sign in equation (5.36). This leads to convective damping for waves with outgoing group velocity. The mode which is easiest to destabilize is $n = 0$. For this mode the solution is [5.12]:

$$\omega=\omega_{*e}\left(1-k_y^2\rho^2\right)\left(1-\mathrm{i}\frac{L_n}{L_s}\right). \tag{5.38}$$

This case corresponds to

$$\phi=\Gamma\,\mathrm{e}^{-\mathrm{i}\xi^2/2} \tag{5.39}$$

where Γ is a constant. As we found previously, drift waves have the strongest tendency for instability for small $k_\parallel$ where the electron shielding is inefficient. We thus expect drift waves to be generated near $k_\parallel = 0$ and then propagate towards larger x. When $k_\parallel$ has grown so that $k_\parallel v_{\mathrm{th}\,i} = \omega$, the ion–Landau damping sets in and absorbs the wave, thus preventing reflection at the plasma boundary and justifying the outgoing boundary condition. The extent of the wave in the x direction, due to the limiting effect of ion–Landau damping, can be estimated as

$$\lambda_x=\frac{v_{*e}}{v_{\mathrm{th}\,i}}L_s.$$

In order to have an absolute instability the growth rate of a drift instability must exceed the damping due to convection. It has recently been shown, both analytically and numerically, that if fully kinetic models are used, both the collisional and the universal drift instabilities are only convective. In toroidal systems with poloidal variation of B_0, however, toroidal couplings may introduce absolute instability.

5.4 Interchange Perturbations Analysed by the Energy Principle Method

As mentioned in the first section, one common method of determining the stability of a system is by calculating the change in energy caused by a small perturbation. We shall here apply this method to an exchange of flux tubes (a tube where no magnetic field lines are crossing the 'side' surfaces). Since the most unstable perturbations are electrostatic (no bending of field lines, $k_\parallel = 0$) we shall consider only electrostatic perturbations.

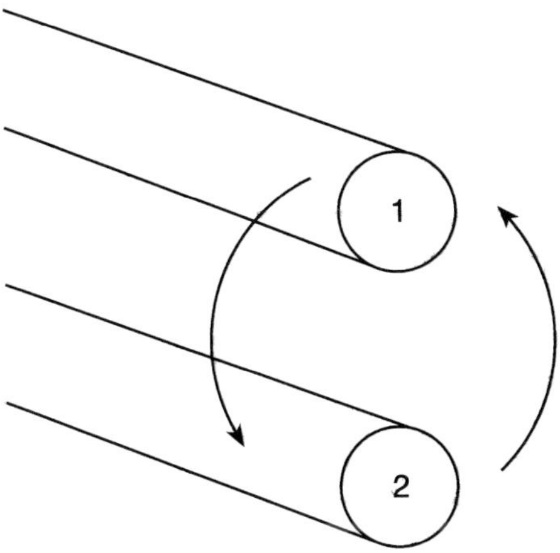

Figure 5.2. Interchange of flux tubes.

For a Maxwellian velocity distribution the average particle energy is

$$E = \tfrac{1}{2}NT$$

where N is the number of degrees of freedom. The equation of state is written

$$p = C(nm)^{\gamma}$$

where

$$\gamma = \frac{2+N}{N} \qquad \text{or} \qquad N = \frac{2}{\gamma - 1}.$$

Accordingly

$$E = \frac{T}{\gamma - 1} \tag{5.40}$$

and the internal energy in a volume v is

$$W_p = nv\frac{T}{\gamma - 1} = \frac{pv}{\gamma - 1} \tag{5.41}$$

where n is the particle density and $p = nT$ is the pressure.

We shall now consider the exchange of plasma and magnetic flux from volume 1 into volume 2 and vice versa according to figure 5.2. Assuming an adiabatic process

$$\frac{\mathrm{d}}{\mathrm{d}t}(pv^{\gamma}) = 0 \tag{5.42}$$

the change in energy can be written

$$\Delta W_p = \frac{1}{\gamma - 1}\left[p_1\left(\frac{v_1}{v_2}\right)^{\gamma} v_2 + p_2\left(\frac{v_2}{v_1}\right)^{\gamma} v_1 - p_1v_1 - p_2v_2\right] \tag{5.43}$$

where we used the relation (5.42) in the form

$$p_1 v_1^\gamma = p_2 v_2^\gamma.$$

For small perturbations we may write $p_1 = p$, $v_1 = v$, $p_2 = p + \delta p$ and $v_2 = v + \delta v$, where $\delta p \ll p$ and $\delta v \ll v$. Introducing these expressions into equation (5.43) we obtain

$$W_p = \delta p \delta v + \gamma p \frac{\delta v^2}{v}. \tag{5.44}$$

Since the second term is always positive a sufficient condition for stability is

$$\delta p \delta v > 0. \tag{5.45}$$

We may write $\phi = BS$, where S is the surface of the cross-section of the flux tube. Since ϕ is constant along the flux tube we may write

$$\delta v = \delta \int S \, \mathrm{d}l = \phi \delta \int \frac{\mathrm{d}l}{B}.$$

When flux tube 2 is closer to the plasma boundary than tube 1, $\delta p < 0$. Condition (5.45) becomes $\delta v < 0$ or

$$\delta \int \frac{\mathrm{d}l}{B} < 0. \tag{5.46}$$

Condition (5.46) shows that configurations where the magnetic field increases on average towards the plasma boundary are stable to flute perturbations ($k_\parallel = 0$). Such configurations are denoted 'average minimum B systems'. For the simple case when B can be approximated by a vacuum field (low β) generated by an external current, we have

$$B = \frac{2\mu_0 I}{R_c}$$

and the condition (5.46) takes the form

$$\delta \int R_c \, \mathrm{d}l < 0 \tag{5.47}$$

showing that the plasma is stable when δR_c on average is negative, corresponding to a generating current situated in the direction of decreasing density. In this situation the magnetic field lines are concave into the plasma.

In the opposite case the contribution $\delta p \delta v$ to $\delta W p$ is destabilizing. In practice it turns out that such a system is normally unstable at least close to the boundary where p is small and, accordingly, also the second term in equation (5.44).

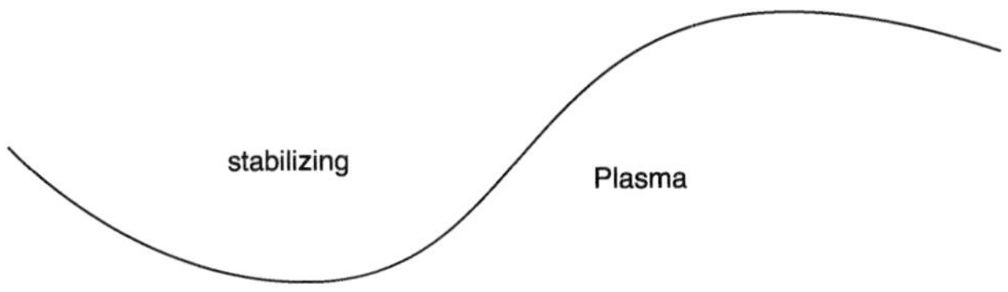

Figure 5.3. Stabilizing and destabilizing curvature regions.

This interchange instability is equivalent to the instability described in section 3.2.1, since the curvature of the magnetic field lines causes a centrifugal force that can be represented as an equivalent gravity

$$g = \frac{v_{\text{th}}^2}{R_c}.$$

We then realize that when the curvature is destabilizing, the gravity will be directed opposite to the density gradient corresponding to the necessary condition for instability $\kappa g > 0$ in equation (3.26) (figure 5.3). Finally, we emphasize once more that the condition (5.46) only says something about the average of the curvature along the field line. A real perturbation in a magnetic confinement device will experience a weighted average of the curvature, which is determined by the mode structure, and only if $k_\parallel = 0$ (flute mode) will the effective curvature be equal to the unweighted curvature, giving the condition (5.46). Finite $k_\parallel$ modes will tend to become trapped in the destabilizing regions, leading to a more unstable situation.

5.5 Eigenvalue Equations for MHD-Type Modes

Since MHD-type modes are more global than drift modes, a WKB approximation is often not valid and a careful inclusion of geometry is required. Thus, we generally have to solve eigenvalue equations in the detailed geometry. We shall give examples here for simplified geometries, which nevertheless show the main features of the problem at the same time as an analytical description of the geometry being possible.

5.5.1 Stabilization of interchange modes by magnetic shear

As mentioned in chapter 3, the electrostatic approximation for interchange modes has to be abandoned in a system with magnetic shear. We thus start from the description of electromagnetic interchange modes in section 3.2.3, but now replacing the gravity drifts by the diamagnetic drifts. The condition $\nabla \cdot \boldsymbol{j} = 0$ now takes the form

$$e\nabla \cdot [n_0 \boldsymbol{v}_{pi} + n(\boldsymbol{v}_{*i} - \boldsymbol{v}_{*e})] = -\nabla \cdot (j_\| \boldsymbol{e}_\|) = \frac{1}{\mu_0}\nabla \cdot (\Delta_\perp A_\| \boldsymbol{e}_\|). \tag{5.48}$$

Here we make use of the approximation $E_\| = 0$, leading to

$$A_\| = -\mathrm{i}\frac{1}{\omega}\boldsymbol{e}_\| \cdot \nabla\phi. \tag{5.49}$$

We shall, in the following, use a cylindrical coordinate system as in section 5.2. The magnetic field will be written as

$$\boldsymbol{B} = B_\theta \boldsymbol{e}_\theta + B_\phi \boldsymbol{e}_\phi \tag{5.50}$$

where $\boldsymbol{e}_\theta$ and $\boldsymbol{e}_\phi$ are unit vectors. Using the representation (5.14) for perturbations we find

$$\boldsymbol{e}_\| \cdot \nabla f = \left[in\left(\frac{\nu}{r}B_\theta - \frac{1}{R}B_\phi\right) + \frac{B_\theta}{r\bar{f}}\frac{\partial \bar{f}}{\partial\theta}\right] f. \tag{5.51}$$

Here ν is essentially the rotational transform q, so that on a rational surface

$$\nu \approx \frac{rB_\phi}{RB_\theta}$$

and we obtain

$$\boldsymbol{e}_\| \cdot \nabla f \approx \frac{1}{\nu R}\left(\frac{1}{\bar{f}}\frac{\partial \bar{f}}{\partial\theta}\right) f.$$

Assuming $R \gg r$ we furthermore have

$$\Delta_\perp f = \left[-n^2\left(\int_0^\theta \frac{\partial\nu}{\partial r}\,\mathrm{d}\theta\right)^2 - \frac{1}{r^2}n^2\nu^2\right] f \tag{5.52}$$

where the r dependence of f was neglected, assuming large mode number n.

If we neglect the θ dependence of ν ($\nu = q$), equation (5.52) reduces to

$$\Delta_\perp f = -n^2\left[\left(\frac{\mathrm{d}q}{\mathrm{d}r}\right)^2\theta^2 + \frac{q^2}{r^2}\right] f = -n^2\frac{q^2}{r^2}(1 + s^2\theta^2) f \tag{5.53}$$

where

$$s = \frac{r}{q}\frac{\mathrm{d}q}{\mathrm{d}r}.$$

The operator expression on the right-hand side of equation (5.48) then takes the form

$$\nabla \cdot (\Delta_\perp A_\| \boldsymbol{e}_\|) = -n^2\frac{q^2}{r^2}\frac{1}{q^2R^2}\frac{\partial}{\partial\theta}(1 + s^2\theta^2)\frac{\partial\phi}{\partial\theta}. \tag{5.54}$$

For the divergence of the diamagnetic flux we use equation (5.18). This means that we take into account the lowest order finite β effect from $\delta B_\parallel$, which enters only in this term. We then also have to know δp, which, to the lowest order, can be taken as a convective perturbation, i.e.

$$\delta p = \mathrm{i}\frac{1}{\omega B_0}(\boldsymbol{e}_\parallel \times \nabla\phi)\cdot\nabla P_0 = -\mathrm{i}\frac{1}{\omega B_0}(\boldsymbol{e}_\parallel \times \nabla P_0)\cdot\nabla\phi. \tag{5.55}$$

Expanding in r/R we now find

$$\boldsymbol{e}_\parallel \times \boldsymbol{\kappa} = \frac{1}{B^2}[\boldsymbol{e}_\parallel \times (\boldsymbol{B}\cdot\nabla)\boldsymbol{B}] \approx \frac{1}{R}(\cos\theta\,\boldsymbol{e}_\theta + \sin\theta\,\hat{\boldsymbol{r}} - \delta\boldsymbol{e}_\theta)$$

where δ is an average part that is of higher order in r/R. This expression gives the local curvature of the magnetic field lines entering equation (5.49). The part δ is the only remaining part when the integral is taken over the whole period in θ. We may express ∇f as in $\boldsymbol{k}f$, where

$$\boldsymbol{k} = n\left(\frac{\mathrm{d}q}{\mathrm{d}r}\theta r + \frac{q}{r}\boldsymbol{e}_\theta - \frac{1}{R}\boldsymbol{e}_\phi\right)$$

with the result

$$(\boldsymbol{e}_\parallel \times \boldsymbol{\kappa})\cdot\boldsymbol{k} = \frac{q}{rR}(\cos\theta + s\theta\sin\theta + \delta). \tag{5.56}$$

Now introducing equations (5.53), (5.54), (5.55) and (5.56) into equation (5.48) we obtain the eigenvalue equation

$$\omega^2(1+s^2\theta^2)\phi + k_c^2 v_A^2\frac{\partial}{\partial\theta}(1+s^2\theta^2)\frac{\partial\phi}{\partial\theta} + Dg(\theta)\phi = 0 \tag{5.57}$$

where

$$k_c = \frac{1}{qR} \qquad D = 2\frac{T_e + T_i}{m_i R}\frac{\mathrm{d}\ln P_0}{\mathrm{d}r}$$

and

$$g(\theta) = \cos\theta + s\theta\sin\theta + \delta. \tag{5.58}$$

Equation (5.57) represents the eigenvalue equation for electromagnetic interchange modes in a toroidal system with circular flux surfaces ($\boldsymbol{B}$ is assumed not to have an $\boldsymbol{r}$ component). The average curvature δ is of order r/R and is not given with sufficient accuracy by the above treatment. We shall here just regard it as a constant of order r/R, using expressions derived in the literature for various systems. As explained in section 5.2, the relevant boundary condition for equation (5.57) in toroidal geometry is $\phi \to 0$ as $\theta \to \infty$. The interchange mode, which we shall consider first as a highly elongated mode, will experience only the average curvature δ in equation (4.51). A common transformation for simplifying equation (5.57) is

$$\varphi = (1+s^2\theta^2)^{1/2}\phi \tag{5.59}$$

leading to the eigenvalue equation

$$\frac{\partial^2\varphi}{\partial\theta^2}+\left[\Omega^2-\frac{s^2}{(1+s^2\theta^2)^2}-\frac{\alpha\delta}{1+s^2\theta^2}\right]\varphi=0 \tag{5.60}$$

where $\Omega = \omega/k_c v_A$ and $\alpha = D/k_c^2 v_A^2$.

Approximate solutions of (5.60) can be obtained by substituting trial functions into the quadratic form

$$\int_0^\infty\left[\varphi\frac{\partial^2\varphi}{\partial\theta^2}+\left(\Omega^2-\frac{s^2}{(1+s^2\theta^2)^2}-\frac{\alpha\delta}{1+s^2\theta^2}\right)\varphi^2\right]\mathrm{d}\theta=0. \tag{5.61}$$

The asymptotic solution to equation (5.60) for $\Omega^2 = 0$ is

$$\varphi=(1+s^2\theta^2)^{1/2}(A\theta^{\gamma_1}+B\theta^{\gamma_2})$$

where

$$\gamma_{1,2}=-\tfrac{1}{2}[1\pm(1+4\alpha\delta/s^2)^{1/2}].$$

It can be shown that the probability of smoothly connecting this solution to the region $\theta \approx 0$, and at the same time making $\Omega^2 < 0$ in equation (5.55), depends on the sign of $1 + 4\alpha\delta/s^2$ giving stability when this expression is positive. The stability condition is thus

$$\tfrac{1}{4}s^2+\alpha\delta>0. \tag{5.62}$$

Since $\alpha = -q^2 R\,\mathrm{d}\beta/\mathrm{d}r$ we now obtain the Mercier criterion (3.32) for $\delta = (r/R)(1 - 1/q^2)$, and the Suydam criterion (3.31) for $\delta = -(r/R)q^2$, corresponding to toroidal and cylindrical geometries, respectively.

5.5.2 Ballooning modes

Another type of solution to equation (5.57) is a mode that varies strongly on the $\cos\theta$ space scale. Such a mode may localize in regions where the normal curvature $\cos\theta > 0$, thus experiencing unfavourable curvature on average. For $s \sim 1$ it turns out that the $s\theta\sin\theta$ part of the curvature (named geodesic curvature) substantially extends the unfavourable curvature region. This is a ballooning mode. Since $\delta \sim a/R$ we shall here neglect the average curvature. As it turns out, ballooning modes are also very sensitive to a θ dependence of ν, which is the lowest order effect in β of a deviation from circular flux surfaces. If we assume a harmonic variation of ν with θ, i.e.

$$\nu(r,\theta)=q(r)+\bar{\nu}\cos\theta$$

we have to modify equation (5.53) to

$$\Delta_\perp f=-n^2\frac{q^2}{r^2}\left[1+\left(s\theta-\frac{\bar{\nu}}{q}\sin\theta\right)^2\right]f \tag{5.63}$$

and equation (5.54) accordingly.

We also find that equation (5.56) is replaced by

$$(\boldsymbol{e}_{\parallel} \times \boldsymbol{\kappa}) \cdot \boldsymbol{k} = \frac{q}{rR}\left[\cos\theta + \left(s\theta - \frac{\nu}{q}\sin\theta\right)\sin\theta\right]. \tag{5.64}$$

As can be shown by analytical solutions for the equilibrium at small β, we have the relation $\bar{\nu}/q = \alpha$. The eigenvalue equation for ballooning modes then takes the form

$$[1 + (s\theta - \alpha\sin\theta)^2]\Omega^2\phi + \frac{\partial}{\partial\theta}[1 + (s\theta - \alpha\sin\theta)^2]\frac{\partial\phi}{\partial\theta} + \alpha g(\theta)\phi = 0 \tag{5.65}$$

where

$$g(\theta) = g^{(1)}(\theta) + g^{(2)}(\theta) \tag{5.66}$$

$$g^{(1)}(\theta) = \cos\theta + s\theta\sin\theta \tag{5.67}$$

$$g^{(2)}(\theta) = -\alpha\sin^2\theta. \tag{5.68}$$

The eigenvalue equation (5.65) can be solved analytically for small s by deriving a quadratic form, and using a trial function derived for small s and α by symmetric expansion. We again introduce the transformation (5.59). The lowest order eigenfunction can then be obtained by ignoring the slow $s\theta$ dependence as

$$\varphi^{(1)} = \frac{\alpha g^{(1)}}{1 + s^2\theta^2}\langle\varphi\rangle \tag{5.69}$$

where $\langle\varphi\rangle$ is a slow background variation and Ω^2 is assumed to be small. The next order φ that enters is the second harmonic part

$$\varphi^{(2)} = \frac{\alpha^2}{4(1 + s^2\theta^2)^2}[(1 - 2s^2\theta^2)\cos 2\theta + 3s\theta\sin 2\theta]\langle\varphi\rangle. \tag{5.70}$$

The average φ, $\langle\varphi\rangle$, asymptotically has to take the form $\langle\varphi\rangle \sim \mathrm{e}^{\pm\mathrm{i}\Omega\theta}$, which is the same as would be obtained from equation (5.60). In the inner region we may, for ballooning modes at small s, make a constant approximation. The transition is estimated to occur at $\theta \sim 1/s$. We thus take the ansatz for $\langle\varphi\rangle$ as

$$\langle\varphi\rangle \simeq \begin{cases} 1 & \theta \le \kappa/s \\ \mathrm{e}^{\mathrm{i}\Omega(\theta - \kappa/s)} & \theta \ge \kappa/s \end{cases} \tag{5.71}$$

where κ can in principle be determined by maximizing the growth rate variationally.

The result obtained from integrating a variational form in φ, corresponding to equation (5.61), and using

$$\varphi = \langle\varphi\rangle + \varphi^{(1)} + \varphi^{(2)}$$

can be written in the form

$$\mathrm{i}(1+a)\Omega + \left(\frac{\kappa}{s} + b\right)\Omega^2 = \delta W \tag{5.72}$$

where δW is the energy change in dimensionless form given by

$$\delta W = \frac{\pi}{4s}\left[s^2 - \frac{3}{2}\alpha^2 s + \frac{9}{32}\alpha^4 - \frac{5}{2}\alpha\, \mathrm{e}^{-1/s}\right]. \tag{5.73}$$

Here the last term in equation (5.73) is due to a mixing of space scales $s\theta$ and $\cos\theta$ in the integration, while the constants a and b are due to the overlapping of the space scales $s\theta$ and $\mathrm{i}\Omega\theta$. Since a and b depend on κ in a rather complicated way a variational determination of κ is not practical. The solution for $\langle\varphi\rangle$ can, in principle, be obtained from the 'averaged' equation

$$\frac{\partial^2\langle\varphi\rangle}{\partial\theta^2} + \left[\Omega^2 - \frac{s^2}{(1+s^2\theta^2)^2} + \frac{2\alpha^2 s - (3/8)\alpha^4}{(1+s^2\theta^2)^3}\right]\langle\varphi\rangle = 0. \tag{5.74}$$

As can be verified numerically, the asymptotic solution $\mathrm{e}^{\mathrm{i}\Omega(\theta-\kappa/s)}$ holds essentially from the point of inflection for $\langle\varphi\rangle$. This is in the centre of the unstable region given by $\alpha^2 \approx 8/3$, $\Omega^2 \approx -\pi^2/16s^2$ and $\theta_{\mathrm{infl}} \approx 0.5/s$, where θ_{infl} is the inflection point. Now, continuing the asymptotic solution to smaller θ, we realize that it will reach 1 somewhere in the interval $0 \le \theta \le \theta_{\mathrm{infl}}$. The simplest possible choice is then to take the matching point in the middle of this interval, i.e.

$$\frac{\kappa}{s} = \frac{1}{2}\theta_{\mathrm{infl}} = \frac{1}{4s} \tag{5.75}$$

or $\kappa = 0.25$. This value gives good agreement between numerical and analytical results for small s. For $\kappa = 0.25$ we obtain

$$a \simeq \frac{\pi}{4}(-0.69 + 0.57\alpha^2/s - 0.11\alpha^4/s^2) \tag{5.76}$$

$$b \simeq \frac{\pi}{4s}(0.3 - 0.18\alpha^2/s + 0.03\alpha^4/s^2). \tag{5.77}$$

As it turns out, the solution of equation (5.72) is rather insensitive to the constants a and b, apparently due to cancellation effects. The $\mathrm{i}\Omega$ term in equation (5.72) is due to convective damping and represents the most important part of the frequency dependence in equation (5.72). When all terms are included, extremely good agreement is obtained between the growth rate obtained from equation (5.72) and numerical results for small s. The agreement is, however, still within 20% in the centre of the unstable region for $s \sim 0.25$. The stability boundary as given by $\delta W = 0$ is shown in figure 5.4. We note the presence of the two stability regions: one for small α and one for large α. The stability for large α is due to the θ dependence of ν. It is due to a reduction of the geodesic curvature due to finite pressure modification of the equilibrium. When

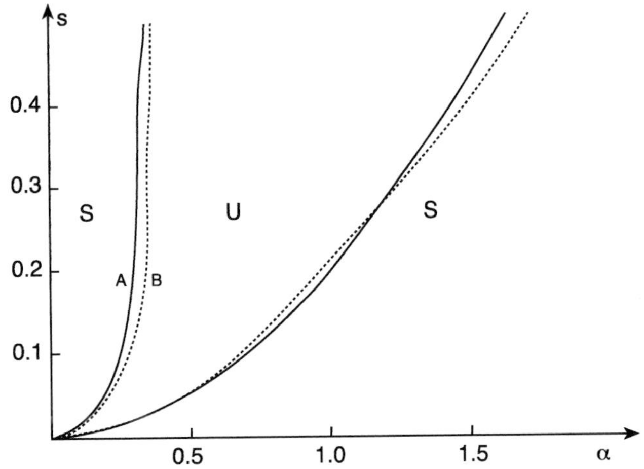

Figure 5.4. Stability boundaries of the MHD ballooning mode. A, numerical; B, analytical. (From Andersson P and Weiland J 1986 *Phys. Fluids* **29** 1744, courtesy of the American Institute of Physics.)

the length of the destabilizing region decreases, the electromagnetic restoring force, through $k_\parallel$, has to increase.

If we include the lowest order FLR effect in a way corresponding to equation (3.44), we can simply make the substitution $\Omega^2 \longrightarrow \Omega(\Omega - \Omega_{*i})$. In this case the convective damping also influences the stability condition, which takes the form

$$\frac{1}{2}|\Omega_{*i}|(1+a) + \frac{1}{4}\left(\frac{\kappa}{s} + b\right)\Omega_{*i}^2 + \delta W > 0. \tag{5.78}$$

This condition should be compared to the condition (3.46) in the shearless case.

5.5.3 Kink modes

While the interchange mode can be unstable inside the plasma (internal mode), the kink mode is more or less associated with the plasma boundary. It is due to a plasma current with a transverse gradient and can, in the slab description, easily be included, as shown in section 3.2.4. For a current profile extending over the whole cross-section, it leads to a driving term $\Omega_{ci} k_\parallel v_0/n$, where v_0 is due to the background current and n is the toroidal mode number. As a starting point we shall consider the simple pinch in figure 5.5.

When the cross-section of the current is decreased the magnetic field pressure increases and enhances the perturbation. For a toroidal configuration the pinch instability corresponds to $m = 0$ (no poloidal variation). In a system

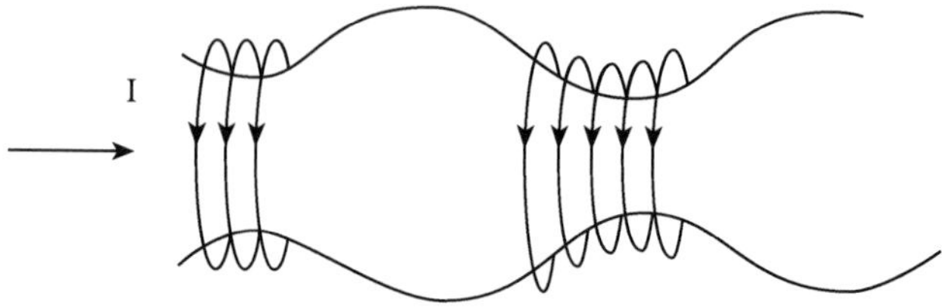

Figure 5.5. The sausage instability corresponding to an $m = 0$ mode in a torus.

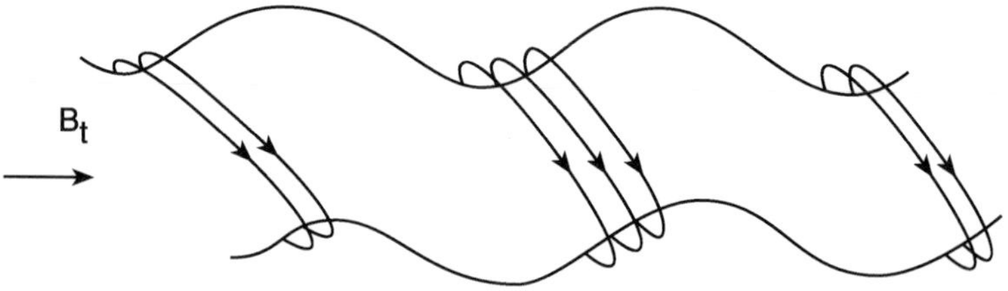

Figure 5.6. A kink perturbation with $m/n = B_t/B_p$ in a torus.

with toroidal magnetic field (along I in figure 5.5), the simple pinch instability is counteracted by the bending of the toroidal magnetic field lines. In a new configuration the total magnetic field will wind around the plasma in the way shown in figure 5.6. This system is now unstable to the perturbation shown in figure 5.6. Here we can see that the instability occurs again in such a way that a bending of the magnetic field lines is avoided. The new perturbation, however, has a finite poloidal variation determined by the relative magnitude of the poloidal and toroidal magnetic fields, B_p and B_t, respectively. This variation can be expressed by

$$k_p = k_t \frac{B_t}{B_p}. \tag{5.79}$$

Modes with a slower poloidal variation are stabilized by the toroidal magnetic field, while modes with a more rapid poloidal variation corresponding to a bending of the plasma current in the case shown in figure 5.6, can still occur. For a toroidal machine with $k_p = m/r$ and $k_t = n/R$, where r is the small radius and R is the large radius, the condition (5.79) becomes (compare section 5.2)

$$\frac{m}{r} = \frac{n}{R}\frac{B_t}{B_p}$$

while the stability criterion $k_p < k_t B_t/B_p$ becomes

$$q > \frac{m}{n}. \tag{5.80}$$

Since $n \geq 1$ for kink modes, a sufficient condition for stability against mode m can be written as $q > m$. For $m = 1$ this condition reduces to the Kruskal–Shafranov limit. The mode $m = 1$ is the least localized mode and extends over most of the cross-section. It also has the largest growth rate, which, for a parabolic current profile, can become a considerable fraction of the Alfvén frequency for ballooning modes v_A/qR. For larger m, the kink modes become more and more localized to the plasma boundary.

We shall now make a more quantitative analysis of the kink mode using a cylindrical geometry. This means that we use the representation (5.4)

$$f(r, \theta, \phi) = \bar{f}(r)\, \mathrm{e}^{\mathrm{i}(m\theta - n\phi)}$$

neglecting the background inhomogeneity of the system in the θ direction. In this case the operators take the form

$$\nabla = \hat{\boldsymbol{r}}\frac{\partial}{\partial r} + \mathrm{i}\frac{m}{r}\boldsymbol{e}_\theta - \mathrm{i}\frac{n}{R}\hat{\boldsymbol{e}}_\phi \tag{5.81}$$

$$\Delta = \frac{1}{r}\frac{\partial}{\partial r}r\frac{\partial}{\partial r} - \frac{m^2}{r^2} - \frac{n^2}{R^2} \tag{5.82}$$

$$\boldsymbol{e}_\parallel \cdot \nabla = \mathrm{i}k_\parallel(r) = \mathrm{i}k_c(m - nq). \tag{5.83}$$

We shall, in the following, for brevity use the symbol $k_\parallel$ for $k_c(m-nq)$, keeping in mind its dependence on r. As mentioned in chapter 3, we may neglect the density perturbation from the polarization drift in equation (3.34). This equation is written in a general operator form, and all we have to do is replace $\hat{\boldsymbol{z}}$ with a space-dependent $\boldsymbol{e}_\parallel$, and use the operator expression (5.75). We shall, however, also replace the gravity drifts by a real curvature, i.e., diamagnetic drifts with space-dependent B_0 and $\boldsymbol{e}_\parallel$. This leads to

$$\nabla \cdot n(\boldsymbol{v}_{*i} - \boldsymbol{v}_{*e}) = \frac{2}{eB_0}(\boldsymbol{e}_\parallel \times \boldsymbol{\kappa}) \cdot \nabla\delta P$$

where

$$\boldsymbol{\kappa} = (\boldsymbol{e}_\parallel \cdot \nabla)\boldsymbol{e}_\parallel = -\frac{\boldsymbol{R}_c}{R_c^2}.$$

Then, using a convective pressure perturbation

$$\delta P = -\boldsymbol{\xi}_\perp \cdot \nabla P_0$$

where $\boldsymbol{\xi}_\perp = -\frac{\mathrm{i}}{\omega B_0}(\boldsymbol{e}_\parallel \times \nabla\phi)$ is the displacement, leads to the result

$$\nabla \cdot n(\boldsymbol{v}_{*i} - \boldsymbol{v}_{*e}) = 2\mathrm{i}(m/r)^2 \frac{1}{\omega e B_0^2 R_c}\frac{\mathrm{d}P_0}{\mathrm{d}r}\phi. \tag{5.84}$$

We then arrive at the eigenvalue equation

$$\omega^2 \frac{1}{r}\frac{\partial}{\partial r}r\frac{\partial\phi}{\partial r} - \left(\frac{m^2}{r^2} + \frac{n^2}{R^2}\right)(\omega^2 - k_\parallel^2 v_A^2)\phi = k_\parallel v_A \frac{1}{r}\frac{\partial}{\partial r}r\frac{\partial}{\partial r}k_\parallel v_A \phi$$
$$- \frac{B_0}{n_0 m_i}\frac{\mathrm{d}J_\parallel}{\mathrm{d}r}k_\parallel \frac{m}{r}\phi - 2(m/r)^2 \frac{1}{m_i n R_c}\frac{\mathrm{d}P_0}{\mathrm{d}r}\phi. \tag{5.85}$$

The question of stability is most easily studied in the energy integral formulation. Thus, multiplying equation (5.85) by $r\phi$ and integrating from $r = 0$ to a, performing partial integrations of the terms containing $(\partial/\partial r)r(\partial/\partial r)$, we obtain the energy formulation

$$\omega^2 \int_0^a r\left(\frac{\mathrm{d}\phi}{\mathrm{d}r}\right)^2 \mathrm{d}r + \int_0^a (\omega^2 - k_\parallel^2 v_A^2)\frac{m^2}{r^2}\phi^2 r\,\mathrm{d}r - v_A^2 \int_0^a \left(\frac{\mathrm{d}}{\mathrm{d}r}k_\parallel\phi\right)^2 r\,\mathrm{d}r - \frac{B_0}{m_i n}\int_0^a m\frac{\mathrm{d}J_\parallel}{\mathrm{d}r}k_\parallel\phi^2\,\mathrm{d}r = \left(\phi a\frac{\mathrm{d}\phi}{\mathrm{d}r} - k_\parallel v_A\phi r\frac{\mathrm{d}}{\mathrm{d}r}k_\parallel v_A\phi\right)_{r=a} \tag{5.86}$$

where we neglected terms of order r/R. Since now

$$\int_0^a \left(\frac{\mathrm{d}}{\mathrm{d}r}k_\parallel\phi\right)^2 r\,\mathrm{d}r = \int_0^a \left[k_\parallel^2\left(\frac{\mathrm{d}\phi}{\mathrm{d}r}\right)^2 + \left(\frac{\mathrm{d}k_\parallel}{\mathrm{d}r}\right)^2\phi^2 - \frac{1}{2}\phi^2\frac{1}{r}\frac{\mathrm{d}}{\mathrm{d}r}\left(r\frac{\mathrm{d}k_\parallel^2}{\mathrm{d}r}\right)\right] r\,\mathrm{d}r - \frac{1}{2}\left(a\frac{\mathrm{d}k_\parallel^2}{\mathrm{d}r}\phi^2\right)_{r=a}$$

we may write equation (5.86) for $\phi(a) = 0$ (internal mode) as

$$\int_0^a \left[f(r)\left(\frac{\mathrm{d}\phi}{\mathrm{d}r}\right)^2 + g(r)\phi^2\right]\mathrm{d}r = 0 \tag{5.87}$$

where

$$f(r) = (k_\parallel^2 v_A^2 - \omega^2)r \tag{5.88}$$

and

$$g(r) = (k_\parallel^2 v_A^2 - \omega^2)\frac{m^2}{r} + v_A^2\left(\frac{\mathrm{d}k_\parallel}{\mathrm{d}r}\right)^2 r - \frac{1}{2}v_A^2\frac{\mathrm{d}}{\mathrm{d}r}\left(r\frac{\mathrm{d}k_\parallel^2}{\mathrm{d}r}\right) + \frac{B_0}{m_i n}\frac{\mathrm{d}J_\parallel}{\mathrm{d}r}mk_\parallel. \tag{5.89}$$

If the terms proportional to ω^2 are separated out, the remaining terms are proportional to the energy change δW ($\omega^2 < 0$ corresponds to the unstable case where $\delta W < 0$). The expression (5.89) for $g(r)$ can be simplified if we make use of the relation between magnetic shear $\mathrm{d}k_\parallel/\mathrm{d}r$ and current. To lowest order in the inverse aspect ratio, the ϕ component of Ampère's law may be written

$$\frac{1}{r}\frac{1}{2}\frac{\mathrm{d}}{\mathrm{d}r}(rB_\theta) = \mu_0 J_\phi \approx \mu_0 J_\parallel \tag{5.90}$$

while

$$\frac{\mathrm{d}k_\parallel}{\mathrm{d}r} = m\frac{\mathrm{d}k_c}{\mathrm{d}r} = -mk_c\left(\frac{1}{r} - \frac{1}{B_\theta}\frac{\mathrm{d}B_\theta}{\mathrm{d}r}\right). \tag{5.91}$$

Combining equations (5.90) and (5.91) we obtain

$$J_{\parallel} = \frac{1}{\mu_0} B_\theta \left(qR \frac{\mathrm{d}k_c}{\mathrm{d}r} + \frac{2}{r} \right)$$

and accordingly

$$\frac{B_0 m}{n_0 m_i} \frac{\mathrm{d}J_{\parallel}}{\mathrm{d}r} = v_A^2 \left(r \frac{\mathrm{d}^2 k_{\parallel}}{\mathrm{d}r^2} + 3 \frac{\mathrm{d}k_{\parallel}}{\mathrm{d}r} \right). \tag{5.92}$$

We then obtain

$$g(r) = \frac{m^2}{r} (\omega^2 - k_{\parallel}^2 v_A^2) - v_A^2 \frac{\mathrm{d}k_{\parallel}^2}{\mathrm{d}r}. \tag{5.93}$$

Another simplification is obtained if we change the dynamic variable to $\xi = \phi/r$ (the radial component of the plasma displacement is $\xi_r = 1/(\omega B)(m/r)\phi$). We then obtain the energy formulation

$$\int_0^a \left[f(r) \left(\frac{\mathrm{d}\xi}{\mathrm{d}r} \right)^2 + h(r)\xi^2 \right] \mathrm{d}r = 0 \tag{5.94}$$

where

$$f(r) = r^3 (k_{\parallel}^2 v_A^2 - \omega^2)$$

and

$$h(r) = (m^2 - 1)(k_{\parallel}^2 v_A^2 - \omega^2) r.$$

For stability we require $\omega^2 > 0$, and obtain the condition

$$\int_0^a (m - nq)^2 \left[\left(r \frac{\mathrm{d}\xi}{\mathrm{d}r} \right)^2 + (m^2 - 1)\xi^2 \right] r \, \mathrm{d}r > 0. \tag{5.95}$$

This condition is fulfilled for modes with $m > 1$. If $m = 1$, a marginally stable mode with $\mathrm{d}\xi/\mathrm{d}r = 0$ can be constructed if $q(0) < 1$. In this case, higher order terms in r/R have to be included in order to determine stability.

For external modes, $\phi(a) \neq 0$, appropriate boundary conditions have to be imposed at the plasma boundary. These are the conditions of pressure balance across the surface

$$\boldsymbol{B}_0 \cdot (\delta \boldsymbol{B} + \boldsymbol{\xi}_r \cdot \nabla \boldsymbol{B}_0) = \text{constant}$$

and the condition that the displaced plasma surface remains a flux surface

$$\delta B_r = [\nabla \times (\boldsymbol{\xi}_r \times \boldsymbol{B}_0)]_r$$

where $\boldsymbol{\xi}_r$ is the radial displacement.

If there is no stabilization due to a conducting wall, this leads to the stability condition

$$\int_0^a \left(\frac{1}{q} - \frac{n}{m}\right)^2 \left[\left(r\frac{\mathrm{d}\xi}{\mathrm{d}r}\right)^2 + (m^2 - 1)\xi^2\right] r\,\mathrm{d}r$$
$$+ \left[\frac{2}{q_a}\left(\frac{n}{m} - \frac{1}{q_a}\right) + (1+m)\left(\frac{n}{m} - \frac{1}{q_a}\right)^2\right] a^2\xi_a^2 > 0 \qquad (5.96)$$

where the index a indicates the value at $r = a$.

The condition (5.96) can be violated only if $nq_a < m$, i.e., a condition equivalent to equation (5.80) evaluated at the plasma boundary.

Another way of writing the stability condition is by using the relation (5.90) in the other direction, i.e., expressing all effects of magnetic shear in $J_\parallel$. This leads to the condition

$$\int_0^a \left[\frac{1}{\mu_0}\delta B^2 + B_\theta\left(1 - \frac{nq}{m}\right)\frac{\mathrm{d}J_\parallel}{\mathrm{d}r}\xi_r^2\right] r\,\mathrm{d}r > 0 \qquad (5.97)$$

which shows that for $\mathrm{d}J_\parallel/\mathrm{d}r < 0$, which is the usual case, only the regions where $nq < m$ are destabilizing.

Finally, we also note that we can use equation (5.92) to rewrite equation (5.85) in the form

$$\frac{\partial}{\partial r}\left[(k_\parallel^2 v_A^2 - \omega^2) r\frac{\partial\phi}{\partial r}\right] + \frac{m^2}{r^2}(\omega^2 - k_\parallel^2 v_A^2)\phi - v_A^2\frac{\mathrm{d}k_\parallel^2}{\mathrm{d}r^2}\phi + \frac{m^2}{r}\frac{2}{m_i n_0 R}\frac{\mathrm{d}P_0}{\mathrm{d}r}\phi = 0. \qquad (5.98)$$

Equation (5.98) agrees with equation (19*a*) in [5.38] if the displacement $\xi \sim \phi/r$ is introduced. An important property of equation (5.98) is the presence of singularities when $k_\parallel^2 v_A^2 = \omega^2$. Assuming that near such a singularity $\partial\phi/\partial r \gg \phi/r$ we may neglect all terms except the first in equation (5.98), this term can then be integrated to

$$\frac{\partial\phi}{\partial r} = \frac{C}{r(k_\parallel^2 v_A^2 - \omega^2)}$$

where C is a constant of integration, thus justifying our WKB approximation close to $k_\parallel^2 v_A^2 = \omega^2$, and showing the presence of a singularity. Since $k_\parallel v_A$ is usually a monotonic function of r, we may have solutions in a continuous range of ω with the location of the singularity varying with ω. This continuous range of solutions is usually referred to as the *Alfvén continuum*. In the presence of toroidicity there will, however, exist a minimum in $k_\parallel^2 v_A^2$. Then, in the region where ω^2 is smaller than this minimum, there is no singularity and the eigenvalues of ω form a discrete spectrum. These modes are referred to as *global modes* since they are not restricted in space by a singularity.

5.6 The Drift Kinetic Equation

The complexity of a full Vlasov description in a magnetized plasma has led to the development of a number of simplified approximate descriptions in various limits. One obvious limit is the case of strongly magnetized particles [5.16, 5.20]. In this limit the particles are well localized in the plane perpendicular to the magnetic field so that the kinetic description is needed only along the field. This approximation is usually valid for electrons in laboratory plasmas and sometimes also for ions. The condition for localization in the perpendicular plane may be written $\rho \ll \lambda$, where ρ is the Larmor radius and λ is the inhomogeneity scale length of the phenomenon we wish to study. Another simplification that still leaves a large class of important phenomena within the description is to assume that the time scale of the phenomenon we are interested in is much longer than the gyroperiod, i.e., $\omega \ll \Omega_{ci}$. When these conditions are fulfilled it is easy to average the Vlasov equation over a Larmor orbit, since both the distribution function and the electromagnetic fields are almost constants over the Larmor orbit in this case. We shall, however, include magnetic curvature, thus keeping background drifts proportional to ρ/L_B, where L_B is the inhomogeneity scale length of the background magnetic field. This effect may sometimes be more important than FLR effects of order ρ/λ, since it enters multiplied by the large-scale thermal velocity.

We shall now use a Lagrangian method of averaging, i.e., we follow the particle around the gyro orbit instead of averaging over the short time-scale at a point. The averaging procedure is then considerably simplified since we already know the average of the particle velocity, i.e., the guiding centre drift. We thus have

$$\langle \boldsymbol{v} \rangle = \boldsymbol{v}_{gc} \approx \frac{1}{B_0}(\boldsymbol{E} \times \boldsymbol{e}_\parallel) + v_\parallel \frac{\delta \boldsymbol{B}_\perp}{B_0} + \boldsymbol{v}_D \tag{5.99}$$

where

$$\begin{aligned} \boldsymbol{v}_D &= \boldsymbol{v}_\kappa + \boldsymbol{v}_{\nabla B} \\ \boldsymbol{v}_\kappa &= \frac{v_\parallel^2}{\Omega_c}(\boldsymbol{e}_\parallel \times \boldsymbol{\kappa}) \\ \boldsymbol{v}_{\nabla B} &= \frac{v_\perp^2}{2\Omega_c}(\boldsymbol{e}_\parallel \times \nabla \ln B) \\ \boldsymbol{\kappa} &= \boldsymbol{e}_\parallel \cdot \nabla \boldsymbol{e}_\parallel . \end{aligned}$$

$\boldsymbol{E}$ is the electric field, B_0 is the background magnetic field and $\delta \boldsymbol{B}_\perp$ is the perpendicular magnetic field perturbation.

We start from the Vlasov equation in the form

$$\frac{\partial f}{\partial t} + \boldsymbol{v} \cdot \nabla f + \frac{q}{m}(\boldsymbol{E} + \boldsymbol{v} \times \boldsymbol{B}) \cdot \frac{\partial f}{\partial \boldsymbol{v}} = 0. \tag{5.100}$$

We now separate our description in the directions parallel and perpendicular to $\boldsymbol{B}_0$, using the notations $\parallel$ and $\perp$, respectively. The velocity is then

$$\boldsymbol{v} = \boldsymbol{v}_\perp + v_\parallel \boldsymbol{e}_\parallel$$

where $\boldsymbol{e}_\parallel = \boldsymbol{B}_0/B$, with phase angle $\phi = \boldsymbol{e}_\parallel \times \boldsymbol{v}_\perp$.

The velocity gradient may now be written

$$\frac{\partial}{\partial \boldsymbol{v}} = \hat{\boldsymbol{v}}_\perp \frac{\partial}{\partial v_\perp} + \hat{\boldsymbol{\phi}} \frac{1}{v_\perp} \frac{\partial}{\partial \phi} + \boldsymbol{e}_\parallel \frac{\partial}{\partial v_\parallel}$$

and then

$$(\boldsymbol{v} \times \boldsymbol{B}) \cdot \frac{\partial}{\partial \boldsymbol{v}} = (\boldsymbol{v}_\perp \times \boldsymbol{B}_\parallel) \cdot \hat{\boldsymbol{\phi}} \frac{1}{v_\perp} \frac{\partial}{\partial \phi} + (\boldsymbol{v} \times \delta \boldsymbol{B}_\perp) \cdot \frac{\partial}{\partial \boldsymbol{v}}.$$

Equation (5.100) then reduces to

$$\frac{\partial f}{\partial t} + \boldsymbol{v} \cdot \nabla f + \frac{q}{m}[\boldsymbol{E} + (\boldsymbol{v} \times \delta \boldsymbol{B}_\perp)] \cdot \frac{\partial f}{\partial \boldsymbol{v}} - \frac{B_\parallel}{B_0} \Omega_c \frac{\partial f}{\partial \phi} = 0 \tag{5.101}$$

where the factor $B_\parallel/B$ accounts for perturbations in $\boldsymbol{B}$ parallel to the background field. This factor is always of order 1.

We shall now use the assumption that Ω_c is much larger than any other frequency in (5.101). To lowest order in Ω_c^{-1}, equation (5.101) then leads to the condition

$$\frac{\partial f}{\partial \phi} = 0.$$

This means that as a first approximation we can treat f as independent of ϕ in all terms, except the last term in equation (5.101). However, we also want to keep curvature terms proportional to $\rho = v_\perp/\Omega_c$. These terms are of first order in Ω_c^{-1} so some care is needed in treating them. We shall assume that $f = f(t, \boldsymbol{r}, v_\parallel^2, v_\perp^2, \phi)$. The most important curvature dependence of f enters the separation between $v_\parallel$ and $v_\perp$. If we separate out this additional space dependence we may write

$$\nabla f = \frac{\partial f}{\partial \boldsymbol{r}} + \frac{\partial f}{\partial v_\perp^2} \nabla v_\perp^2 + \frac{\partial f}{\partial v_\parallel^2} \nabla v_\parallel^2$$

or

$$\nabla f = \frac{\partial f}{\partial \boldsymbol{r}} + v_\parallel \nabla(\boldsymbol{e}_\parallel \cdot \boldsymbol{v}) \left(\frac{1}{v_\parallel} \frac{\partial f}{\partial v_\parallel} - \frac{1}{v_\perp} \frac{\partial f}{\partial v_\perp} \right) \tag{5.102}$$

where the space dependence of $\boldsymbol{e}_\parallel$ has been separated out. It can now be shown that

$$\boldsymbol{v} \cdot \nabla(\boldsymbol{e}_\parallel \cdot \boldsymbol{v}) = v_\parallel (\boldsymbol{v} \cdot \boldsymbol{\kappa}) \tag{5.103}$$

and we notice that this curvature term depends on the phase angle ϕ. We now perform the gyroaveraging of equation (5.101). Then $\partial f/\partial t$ is unchanged.

In the second term $v_\parallel \boldsymbol{e}_\parallel \cdot (\partial f/\partial \boldsymbol{r})$ is unchanged, while $\boldsymbol{v}$ is replaced by $\boldsymbol{v}_{gc}$ in the perpendicular part. In the third term we can neglect $\partial f/\partial\phi$. The part containing $\partial f/\partial v_\perp$ reduces to

$$\frac{q}{m}\left[\boldsymbol{E} + v_\parallel(\boldsymbol{e}_\parallel \times \delta\boldsymbol{B}_\perp)\right] \cdot \frac{\boldsymbol{v}_\perp}{v_\perp}\frac{\partial f}{\partial v_\perp}.$$

Here the averaging leads to a replacement of $\boldsymbol{v}_\perp$ by $\boldsymbol{v}_{gc}$. As it turns out, however, the first two parts of $\boldsymbol{v}_{gc}$ do not contribute, due to orthogonality, so we are left with only $\boldsymbol{v}_D$. In the part containing $\partial f/\partial v_\parallel$ we may simply replace $\boldsymbol{v}_\perp$ by $\boldsymbol{v}_{gc}$. The last term in equation (5.101) is a total derivative in ϕ and vanishes, since orbit averaging means integrating one period in ϕ.

Thus writing down our averaged equation directly as we obtain it after orbit averaging we have

$$\begin{aligned}&\frac{\partial f}{\partial t} + (v_\parallel \boldsymbol{e}_\parallel + \boldsymbol{v}_{gc}) \cdot \frac{\partial f}{\partial \boldsymbol{r}} + v_\parallel^2 \boldsymbol{\kappa} \cdot \boldsymbol{v}_{gc}\left(\frac{1}{v_\parallel}\frac{\partial f}{\partial v_\parallel} - \frac{1}{v_\perp}\frac{\partial f}{\partial v_\perp}\right)\\ &\quad + \frac{q}{m}\left[E_\parallel + (\boldsymbol{v}_{gc} \times \delta\boldsymbol{B}_\perp)\cdot \boldsymbol{e}_\parallel\right]\frac{\partial f}{\partial v_\parallel}\\ &\quad + \frac{q}{m}\left[\boldsymbol{E} + v_\parallel(\boldsymbol{e}_\parallel \times \delta\boldsymbol{B}_\perp)\right] \cdot \frac{\boldsymbol{v}_D}{v_\perp}\frac{\partial f}{\partial v_\perp} = 0. \qquad (5.104)\end{aligned}$$

Since now

$$\boldsymbol{\kappa} \cdot \boldsymbol{v}_{gc} = \frac{1}{B}(\boldsymbol{e}_\parallel \times \boldsymbol{\kappa}) \cdot \boldsymbol{E} + \frac{v_\parallel}{B}(\boldsymbol{e}_\parallel \times \boldsymbol{\kappa}) \cdot (\boldsymbol{e}_\parallel \times \delta\boldsymbol{B}_\perp)$$

the third term may be written as

$$\frac{q}{m}\boldsymbol{v}_\kappa \cdot [\boldsymbol{E} + v_\parallel(\boldsymbol{e}_\parallel \times \delta\boldsymbol{B}_\perp)]\left(\frac{1}{v_\parallel}\frac{\partial f}{\partial v_\parallel} - \frac{1}{v_\perp}\frac{\partial f}{\partial v_\perp}\right).$$

We then obtain the drift kinetic equation

$$\begin{aligned}&\frac{\partial f}{\partial t} + (v_\parallel \boldsymbol{e}_\parallel + \boldsymbol{v}_{gc}) \cdot \frac{\partial f}{\partial \boldsymbol{r}} + \frac{q}{m}[E_\parallel + (\boldsymbol{v}_{gc} \times \delta\boldsymbol{B}_\perp)\cdot \boldsymbol{e}_\parallel]\frac{\partial f}{\partial v_\parallel}\\ &\quad + \frac{q}{m}[\boldsymbol{E} + v_\parallel(\boldsymbol{e}_\parallel \times \delta\boldsymbol{B}_\perp)] \cdot \left(\frac{\boldsymbol{v}_{\nabla B}}{v_\perp}\frac{\partial f}{\partial v_\perp} + \frac{\boldsymbol{v}_\kappa}{v_\parallel}\frac{\partial f}{\partial v_\parallel}\right) = 0. \qquad (5.105)\end{aligned}$$

We notice that the first three terms can easily be obtained from a continuity equation for guiding centres (cf equation (4.51)). Equation (5.105) agrees to first order in the inverse aspect ratio with the drift kinetic equation derived by D'Ippolito and Davidson [5.20], except for the presence of the $\boldsymbol{v}_D \cdot \frac{\partial f}{\partial \boldsymbol{r}}$ term in equation (5.105). This term is comparable to the other curvature terms if $f_1/f_0 \sim q\phi/T$ and is usually kept.

If we replace the dependence on $v_\parallel^2$ and $v_\perp^2$ of f by $E = v^2/2$ and $\mu - (\partial\mu/\partial r)r$, where $\mu = v_\perp^2/2B$, and r is the perpendicular direction of ∇B_0, we can also obtain the mirror force terms kept by D'Ippolito and Davidson. The correction $-(\partial\mu/\partial r)r$ of μ is necessary in order to have conservation to lowest order on the Ω_c^{-1} time scale. We then obtain a correction to equation (5.102) of the form $(\partial f/\partial\mu)\nabla_\parallel\mu$, i.e.

$$\frac{1}{2}v_\perp^2\left(\frac{1}{v_\parallel}\frac{\partial f}{\partial v_\parallel} - \frac{1}{v_\perp}\frac{\partial f}{\partial v_\perp}\right)\boldsymbol{e}_\parallel \cdot \nabla \ln B. \tag{5.106}$$

The drift kinetic equation (5.105) has here been obtained in a comparatively simple way. It does not take into account finite Larmor radius effects of the type ρ/λ, but includes the full parallel dynamics, is fully nonlinear and makes no WKB assumption for the space scale of perturbations.

5.6.1 Moment equations

In order to see what fluid motion it corresponds to we shall now take moments of equation (5.105). This procedure is rather complicated in the presence of $\boldsymbol{v}_D$, which in general depends on both $v_\perp$ and $v_\parallel$. For this reason we shall, in the following, neglect curvature effects for simplicity.

The zeroth moment is then

$$\frac{\partial n}{\partial t} + \boldsymbol{e}_\parallel \cdot \nabla(nu_\parallel) + \int \boldsymbol{v}_{gc} \cdot \nabla f \, \mathrm{d}\boldsymbol{v} = 0$$

where u is the fluid velocity. Now inserting equation (5.99), where $\boldsymbol{v}_D = 0$, we obtain the continuity equation

$$\frac{\partial n}{\partial t} + \frac{1}{q}\left[\frac{\partial}{\partial z}j_\parallel + \frac{\delta \boldsymbol{B}_\perp}{B_0}\cdot\nabla j_\parallel\right] + \frac{1}{B_0}(\boldsymbol{E}\times\boldsymbol{e}_\parallel)\cdot\nabla n = 0 \tag{5.107}$$

where $j_\parallel$ is the parallel current. The first parallel moment of equation (5.105) may be written

$$\frac{\partial}{\partial t}(nu_\parallel) + \boldsymbol{e}_\parallel \cdot \nabla \int v_\parallel^2 f \, \mathrm{d}v_\parallel + \int \boldsymbol{v}_{gc}\cdot\nabla f v_\parallel \, \mathrm{d}v_\parallel - \frac{q}{m}[E_\parallel + (\boldsymbol{v}_E \times \delta\boldsymbol{B}_\perp)\cdot\boldsymbol{e}_\parallel]n = 0$$

where

$$\boldsymbol{v}_E = \frac{1}{B}(\boldsymbol{E}_\parallel \times \boldsymbol{e}_\parallel).$$

Now $v_\parallel = u_\parallel + w_\parallel$, where $w_\parallel$ is the thermal random velocity. Thus

$$\int v_\parallel^2 f \, \mathrm{d}v_\parallel = \int u_\parallel^2 f \, \mathrm{d}v_\parallel + \int w_\parallel^2 f \, \mathrm{d}w_\parallel = nu_\parallel^2 + \frac{1}{m}P$$

where P is the pressure. Now substituting equation (5.107) for $\partial n/\partial t$ we obtain

$$n\frac{\partial u_\|}{\partial t} + nu_\|\boldsymbol{e}_\| \cdot \nabla u_\| + n\frac{1}{B_0}(\boldsymbol{E} \times \boldsymbol{e}_\|) \cdot \nabla u_\| + nu_\|\frac{\delta \boldsymbol{B}_\perp}{B_0} \cdot \nabla u_\|$$
$$+ \frac{1}{m}\boldsymbol{e}_\| \cdot \nabla P + \frac{1}{m}\frac{\delta \boldsymbol{B}_\perp}{B_0} \cdot \nabla P - \frac{q}{m}[E_\| + (\boldsymbol{v}_E \times \delta \boldsymbol{B}_\perp) \cdot \boldsymbol{e}_\|]n = 0. \tag{5.108}$$

Equation (5.108) is the parallel equation of motion in the absence of FLR effects. It is important to note here the absence of diamagnetic drifts in the convective part of the time derivative. It is instructive to rewrite equation (5.108) slightly. We may define the perpendicular guiding centre fluid velocity

$$\boldsymbol{u}_{gc} = \frac{1}{B_0}(\boldsymbol{E} \times \boldsymbol{e}_\|) + u_\|\frac{\delta \boldsymbol{B}_\perp}{B_0}.$$

Introducing now the diamagnetic drift velocity

$$\boldsymbol{v}_* = \frac{1}{qnB_0}(\boldsymbol{e}_\| \times \nabla P)$$

we have

$$\frac{q}{m}(\boldsymbol{v}_* \times \delta \boldsymbol{B}_\perp) \cdot \boldsymbol{e}_\| = \frac{q}{m}(\boldsymbol{e}_\| \times \boldsymbol{v}_*) \cdot \delta \boldsymbol{B}_\perp = -\frac{1}{mn}\frac{\delta \boldsymbol{B}_\perp}{B_0} \cdot \nabla P.$$

We may thus rewrite equation (5.108) in the form

$$\frac{\partial u_\|}{\partial t} + u_\|\boldsymbol{e}_\| \cdot \nabla u_\| + \boldsymbol{u}_{gc} \cdot \nabla u_\| = \frac{q}{m}\{E_\| + [(\boldsymbol{v}_E + \boldsymbol{v}_*) \times \delta \boldsymbol{B}_\perp] \cdot \boldsymbol{e}_\|\} - \frac{1}{mn}\boldsymbol{e}_\| \cdot \nabla P. \tag{5.109}$$

Equation (5.109) is the usual parallel equation of motion, where the diamagnetic drift is included in the $\boldsymbol{v} \times \boldsymbol{B}$ term but not in the convective derivatives.

5.6.2 The magnetic drift mode

We now restrict our consideration to the case $\boldsymbol{e}_\| \cdot \nabla = 0$ and linearize. Equation (5.109) is then

$$\frac{\partial u_\|}{\partial t} = -\frac{e}{m}E_\| - \frac{e}{m}(\boldsymbol{v}_* \times \delta \boldsymbol{B}_\perp) \cdot \boldsymbol{e}_\|.$$

Using the representation $\boldsymbol{A} = A\boldsymbol{e}_\|$ and

$$\delta \boldsymbol{B}_\perp = \nabla \times A\boldsymbol{e}_\| = \nabla_\perp A \times \boldsymbol{e}_\| \qquad \boldsymbol{E} = -\nabla\phi - \frac{\partial A}{\partial t}\boldsymbol{e}_\|$$

we obtain

$$-\mathrm{i}\omega u_\| = -\mathrm{i}\omega\frac{e}{m}A - \mathrm{i}\frac{e}{m}\boldsymbol{v}_* \cdot [(\boldsymbol{k} \times \boldsymbol{e}_\|) \times \boldsymbol{e}_\|]A$$

or

$$u_{\|} = \frac{e}{m} A \left(1 - \frac{\omega_{*e}}{\omega}\right) \qquad \omega_{*e} = \boldsymbol{k} \cdot \boldsymbol{v}_*.$$

The parallel current is

$$j_{\|} = \frac{1}{\mu}(\nabla \times \boldsymbol{B}) \cdot \boldsymbol{e}_{\|} = \frac{1}{\mu}[\nabla \times (\nabla_{\perp} A \times \boldsymbol{e}_{\|})] \cdot \boldsymbol{e}_{\|} = -\frac{1}{\mu}\nabla_{\perp} A.$$

Thus

$$-neu_{\|} = -\frac{ne^2}{m} A \left(1 - \frac{\omega_{*e}}{\omega}\right) = \frac{1}{\mu} k_{\perp}^2 A$$

or

$$\omega = \omega_{*e} \left(1 + \frac{k_{\perp}^2 c^2}{\omega_{pe}^2}\right)^{-1}.$$

This is the dispersion relation of the magnetic drift mode (of equation (4.59)). We see here that the inclusion of the diamagnetic drift in the convective derivative would not cause a negligible modification. We must then conclude that this term is cancelled by stress tensor effects.

Transport due to an enhanced thermal equilibrium spectrum of magnetic drift modes is discussed in section 7.2 [5.51].

5.6.3 The tearing mode

In a more realistic geometry with magnetic shear, the condition $k_{\|} = 0$, which was assumed for the magnetic drift mode, cannot be fulfilled everywhere for modes with finite radial extent. We thus have to solve a radial eigenvalue problem. The characteristic property $\phi = 0$ of the magnetic drift mode will then enter as a boundary condition at the rational surface. This type of mode can be driven unstable by collisions and is called the tearing mode [5.4]. It is one of the main modes of resistive MHD. In a shortwave version it is called the microtearing mode [5.57, 5.58].

5.7 The Gyrokinetic Equation

We have seen in chapter 4 how we can obtain general kinetic equations in a slab geometry by integrating along unperturbed orbits. For more general geometries we have, in the previous section, derived a drift kinetic equation valid in the limit $k_{\perp}\rho \ll 1$. Finally, we shall consider the case $k_{\perp}\rho \sim 1$ in complex geometry. Gyroaveraged equations of this type are called gyrokinetic equations. A pioneering work along these lines is that by Rutherford and Frieman [5.11]. A generally used assumption in this type of equation is $\rho/L \ll 1$, where L is an equilibrium scale-length. We shall here derive a linear gyrokinetic equation by a method that is considerably shorter than conventional methods.

We may write the Vlasov equation in the form:

$$\frac{\mathrm{D}f}{\mathrm{D}t} = 0 \tag{5.110}$$

where

$$\frac{\mathrm{D}}{\mathrm{D}t} = \frac{\partial}{\partial t} + \boldsymbol{v} \cdot \nabla + \frac{q}{m}(\boldsymbol{E} + \boldsymbol{v} \times \boldsymbol{B}) \cdot \frac{\partial}{\partial \boldsymbol{v}}. \tag{5.111}$$

We shall assume a solution of the form

$$f(\boldsymbol{r}, \boldsymbol{v}, t) = f_0(\boldsymbol{r}, \boldsymbol{v}) + f_1(\boldsymbol{r}, \boldsymbol{v}, t)$$

where f_0 is the background distribution and f_1 is a perturbation fulfilling

$$f_1 \ll f_0.$$

For simplicity we shall here omit background electric fields. Then the equation for f_0 becomes

$$\boldsymbol{v} \cdot \nabla f_0 + \frac{q}{m}(\boldsymbol{v} \times \boldsymbol{B}_0) \cdot \frac{\partial f_0}{\partial \boldsymbol{v}} = 0. \tag{5.112}$$

Writing the velocity in cylindrical coordinates we have

$$\frac{\partial}{\partial \boldsymbol{v}} = \hat{\boldsymbol{v}}_\perp \frac{\partial}{\partial v_\perp} + \hat{\varphi}\frac{1}{v_\perp}\frac{\partial}{\partial \varphi} + \hat{\boldsymbol{e}}\frac{\partial}{\partial v_\parallel}$$

where $\perp$ and $\parallel$ refer to $\boldsymbol{B}_0$ and φ is the gyrophase. Since $\hat{\varphi} = \hat{\mathbf{e}}_\parallel \times \hat{\boldsymbol{v}}_\perp$ we can rewrite equation (5.112) in the form

$$\boldsymbol{v} \cdot \nabla f_0 = \Omega_c \frac{\partial f_0}{\partial \varphi} \tag{5.113}$$

showing that a gyrophase dependence of f_0 is associated with inhomogeneity.

We may also write equation (5.112) in the form

$$\boldsymbol{v} \cdot \left[\nabla f_0 + \Omega_c\left(\hat{\mathbf{e}}_\parallel \times \frac{\partial f}{\partial \boldsymbol{v}}\right)\right] = 0.$$

Assuming $\hat{\mathbf{e}}_\parallel \cdot \nabla f_0 = 0$, we obtain the solution

$$\frac{\partial}{\partial \boldsymbol{v}} f_0 = \frac{1}{\Omega_c}(\hat{\mathbf{e}}_\parallel \times \nabla f_0) + \hat{\boldsymbol{v}}_\perp \frac{\partial}{\partial v_\perp} f_0 + \hat{\mathbf{e}}_\parallel \frac{\partial}{\partial v_\parallel} f_0. \tag{5.114}$$

To first order we have

$$\frac{\mathrm{D}_0 f_1}{\mathrm{D}t} = -\frac{q}{m}[\boldsymbol{E} + \boldsymbol{v} \times \delta \boldsymbol{B}] \cdot \frac{\partial}{\partial \boldsymbol{v}} f_0 \tag{5.115}$$

where

$$\frac{\mathrm{D}_0}{\mathrm{D}t} = \frac{\partial}{\partial t} + \boldsymbol{v} \cdot \nabla + \frac{q}{m}(\boldsymbol{v} \times \boldsymbol{B}_0) \cdot \frac{\partial}{\partial \boldsymbol{v}}$$

is the operator along the unperturbed orbit. The unperturbed orbit is given by

$$\boldsymbol{v}(t) = \tilde{\boldsymbol{v}}(t) + \boldsymbol{v}_D(t) + v_\parallel \hat{\boldsymbol{e}}_\parallel \tag{5.116}$$

where $\tilde{\boldsymbol{v}}(t)$ is the pure gyromotion as given by equation (4.4), $\boldsymbol{v}_D(t)$ is the magnetic drift which may be time dependent along the orbit and $v_\parallel$ is the velocity along $\boldsymbol{B}_0$. We now invert equation (5.115) as

$$f_1 = -\frac{q}{m}\int_\infty^t [\boldsymbol{E}(r(t')) + \boldsymbol{v}(t') \times \delta\boldsymbol{B}(r(t'))]\frac{\partial}{\partial \boldsymbol{v}} f_0 \,\mathrm{d}t' \tag{5.117}$$

where $r(t')$ is the unperturbed orbit.

Now, considering Fourier harmonics in time and space we obtain

$$f_k = -\frac{q}{m}\int_0^\infty [\boldsymbol{E}_k + \boldsymbol{v} \times \boldsymbol{B}_k]\frac{\partial}{\partial \boldsymbol{v}} f_0 \,\mathrm{e}^{-\mathrm{i}\alpha(\tau)}\,\mathrm{d}\tau \tag{5.118}$$

where

$$\begin{aligned}\alpha(\tau) &= \boldsymbol{k} \cdot [\boldsymbol{r}(t) - \boldsymbol{r}(t')] - \omega\tau \\ &= \frac{k_\perp v_\perp}{\Omega_c}[\sin(\Omega_{ct} + \varphi - \theta) - \sin(\varphi - \theta)] - \int_{t-\tau}^t \tilde{\omega}(t')\,\mathrm{d}t'\end{aligned}$$

and $\tilde{\omega} = \omega - k_\parallel v_\parallel - \boldsymbol{k} \cdot \boldsymbol{v}_D(t)$, $\tau = t - t'$, $\tilde{v}(0) = v_\perp(\cos\varphi, \sin\varphi)$, $\boldsymbol{k}_\perp = k_\perp(\cos\theta, \sin\theta)$.

We now introduce potentials, i.e.

$$\begin{aligned}\boldsymbol{E} &= -\nabla\varphi - \frac{\partial \boldsymbol{A}}{\partial t} \\ \boldsymbol{B} &= \nabla \times \boldsymbol{A}.\end{aligned}$$

For a Maxwellian distribution, equation (5.114) now leads to

$$\boldsymbol{k} \cdot \frac{\partial}{\partial \boldsymbol{v}} f_0 = \mathrm{i}\frac{m}{T}(\boldsymbol{k}_\perp \cdot \boldsymbol{v}_\perp + k_\parallel v_\parallel - \omega_{*f}) \tag{5.119}$$

where $\omega_{*f} = \boldsymbol{k} \cdot \boldsymbol{v}_{*f}$ and $\boldsymbol{v}_{*f} = T/(m\Omega_c)(\hat{\mathrm{e}}_\parallel \times \nabla \ln f_0)$. As we have seen in chapters 3 and 4, $A_\parallel$ is the most important part of $\boldsymbol{A}$. In order to include compressional parts of the magnetic field perturbation, i.e., $\delta B_\parallel$ (equation (5.18)), we now also include an A_r component. This makes our choice of $\boldsymbol{A}$ general since we have the freedom of the gauge condition. We then find

$$\begin{aligned}[\boldsymbol{E}_k + \boldsymbol{v} \times \boldsymbol{B}_k] \cdot \frac{\partial}{\partial \boldsymbol{v}} f_0 &= -\mathrm{i}\frac{m}{T} f_0 \boldsymbol{k}_\perp \cdot \boldsymbol{v}_\perp \phi_k + \mathrm{i}\frac{m}{T}(\omega - \omega_{*f}) f_0 \boldsymbol{A}_k \cdot \boldsymbol{v}_\perp \\ &\quad -\mathrm{i}(k_\parallel \phi_k - \omega A_{k\parallel})\frac{m}{T} f_0 v_\parallel + \mathrm{i}(\phi_k - v_\parallel A_{k\parallel})\omega_{*f}\frac{m}{T} f_0.\end{aligned} \tag{5.120}$$

Since now

$$\frac{\mathrm{d}}{\mathrm{d}\tau}\,\mathrm{e}^{-\mathrm{i}\alpha(\tau)} = \mathrm{i}[\boldsymbol{k}\cdot\boldsymbol{v}_\perp(\tau) + k_\parallel v_\parallel - \omega]\,\mathrm{e}^{-\mathrm{i}\alpha(\tau)} \tag{5.121}$$

we may rewrite equation (5.118) in the form

$$f_k = f_0\frac{q}{T}\int_0^\infty \left[\phi_k \frac{\mathrm{d}}{\mathrm{d}\tau}\,\mathrm{e}^{-\mathrm{i}\alpha(\tau)} + \mathrm{i}(\omega_{*f} - \omega)(\phi_k - v_\parallel A_{k\parallel} - \boldsymbol{A}_k\cdot\boldsymbol{v}_\perp)\,\mathrm{e}^{-\mathrm{i}\alpha(\tau)}\right]\mathrm{d}\tau \tag{5.122}$$

or

$$\begin{aligned} f_k = f_0\frac{q}{T}\phi_k f_0 + \mathrm{i}f_0\frac{q}{T}(\omega_{*f} - \omega)(\phi_k - v_\parallel A_{k\parallel})\int_0^\infty \mathrm{e}^{-\mathrm{i}\alpha(\tau)}\mathrm{d}\tau \\ - \mathrm{i}\frac{q}{T}f_0(\omega_{*f} - \omega)\int_0^\infty \boldsymbol{A}_k\cdot\boldsymbol{v}_\perp\,\mathrm{e}^{-\mathrm{i}\alpha(\tau)}\mathrm{d}\tau. \end{aligned} \tag{5.123}$$

In an inhomogeneous system the orbit integrals in equation (5.123) generally require knowledge of a very complicated orbit. We shall avoid this complication here by assuming the gyroperiod to be much shorter than any other time scale and performing an average over it. Thus, a general orbit integral is written

$$\int_0^\infty G(\tau)\,\mathrm{d}\tau = \sum_0^\infty \Delta\tau\frac{1}{\Delta\tau}\int_\tau^{\tau+\Delta\tau} G(\tau)\,\mathrm{d}\tau \tag{5.124}$$

where $\Delta\tau$ is a gyroperiod and the integrals normalized by $\Delta\tau$ are the local gyroaverages of an arbitrary function $G(\tau)$, subject only to the above assumption of time scales. In the gyroaveraging, we can ignore all variation on time scales longer than $\Delta\tau$. Since the time steps $\Delta\tau$ are small as compared to the longer time scales in the system we can convert the summation back to an integral over the long time scale. Thus

$$\int_0^\infty G(\tau)\,\mathrm{d}\tau = \int_0^\infty \langle G(\tau)\rangle\mathrm{d}\tau.$$

Now since

$$\begin{aligned} \exp[-\mathrm{i}(k_\perp v_\perp/\Omega_c)\sin(\Omega_c\tau + \varphi - \theta)] = \sum_n J_n\left(\frac{k_\perp v_\perp}{\Omega_c}\right) \\ \times\exp[-\mathrm{i}n(\Omega_c\tau + \varphi - \theta)] \end{aligned}$$

we obtain

$$\langle \mathrm{e}^{\mathrm{i}\alpha(\tau)}\rangle = J_0(\xi)\exp[\mathrm{i}\xi\sin(\varphi - \theta)]\exp\left[\mathrm{i}\int_{t-\tau}^t \tilde{\omega}(t')\,\mathrm{d}t'\right] \tag{5.125}$$

where $\xi = k_\perp v_\perp/\Omega_c$. Moreover, writing $\boldsymbol{A}_k \cdot \boldsymbol{v}_\perp = A_\perp v_\perp \cos(\Omega_c \tau + \phi - \theta')$ we have

$$
\begin{aligned}
&\langle \boldsymbol{A}_k \cdot \boldsymbol{v}_\perp \, \mathrm{e}^{-\mathrm{i}\alpha(\tau)} \rangle = A_\perp v_\perp \langle \cos(\Omega_c \tau + \varphi - \theta')\, \mathrm{e}^{-\mathrm{i}\alpha(\tau)} \rangle \\
&= \frac{1}{2} \mathrm{e}^{\mathrm{i}L_k} \Bigg\langle \sum_n \mathrm{e}^{\mathrm{i}n\theta} \Big\{ J_{n+1} \exp[-\mathrm{i}n(\Omega_c \tau + \varphi) + \mathrm{i}(\theta - \theta')] \\
&+ J_{n-1} \exp[\mathrm{i}n(\Omega_c \tau + \varphi) - \mathrm{i}(\theta - \theta')] \Big\} \\
&\times \exp\left[\mathrm{i} \int_{t-\tau}^{t} \tilde{\omega}(t')\, \mathrm{d}t' \right] \Bigg\rangle = \mathrm{i} \frac{v_\perp}{k_\perp} (\hat{\mathbf{e}}_\parallel \times \boldsymbol{k}) \cdot \boldsymbol{A} \frac{\mathrm{d}J_0}{\mathrm{d}\xi} \\
&\times \exp[\mathrm{i}L_k] \exp\left[\mathrm{i} \int_{t-\tau}^{t} \tilde{\omega}(t')\, \mathrm{d}t' \right]
\end{aligned}
\tag{5.126}
$$

where

$$
L_k = \frac{k_\perp v_\perp}{\Omega_c} \sin(\varphi - \theta) = (\boldsymbol{v} \times \hat{\mathbf{e}}_\parallel) \cdot \boldsymbol{k}/\Omega_c .
$$

As mentioned above, the integration on the long timescale requires detailed knowledge of particle orbits. If we instead differentiate with respect to the long timescale we obtain the gyrokinetic equation

$$
(\omega - k_\parallel v_\parallel - \omega_D) g_k = \frac{q}{T} (\omega - \omega_{*f}) [(\phi_k - v_\parallel A_\parallel) J_0(\xi) - \mathrm{i} \frac{v_\perp}{k_\perp} (\hat{\mathbf{e}}_\parallel \times \boldsymbol{k}) \cdot \boldsymbol{A}_k J_0'] \tag{5.127}
$$

where

$$
\begin{aligned}
g_k &= \left(f_k + \frac{q\phi}{T} f_0 \right) \mathrm{e}^{-\mathrm{i}L_k} \\
J_0' &= \frac{\mathrm{d}J_0}{\mathrm{d}\xi} \\
\xi &= \frac{k_\perp v_\perp}{\Omega_c} \\
L_k &= (\boldsymbol{v} \times \hat{\mathbf{e}}_\parallel) \cdot \frac{\boldsymbol{k}}{\Omega_c} .
\end{aligned}
$$

Here $\omega_D = \boldsymbol{k} \cdot \boldsymbol{v}_D(v_\parallel^2, v_\perp)$ as given in the derivation of the drift kinetic equation. The diamagnetic drift frequency contains ∇f_0 and is also velocity dependent. For a Maxwellian distribution, where both n and T are space dependent, we find

$$
\omega_{*f} = \omega_* \left[1 + \eta \left(\frac{mv^2}{2T} - \frac{3}{2} \right) \right] \tag{5.128}
$$

where $\eta = L_n/L_T$ and ω_* is the usual fluid diamagnetic drift with only a density gradient. Equation (5.127) agrees with the gyrokinetic equation obtained by Antonsen and Lane [5.53].

5.7.1 Applications

It is straightforward to re-derive the results on both electrostatic and electromagnetic modes in chapter 4 from equation (5.127). The advantage of equation (5.127) is that it allows for space-dependent coefficients (i.e., ω_*, ω_D, etc). We also note that $k_\parallel$ in general has to be treated as an operator. Another major difference is that equation (5.127) is valid for arbitrary ω_D/ω, while the treatment in chapter 4 only works to first order in ω_D/ω. We shall explore this property a little in the electrostatic limit. We shall also take the limit $\omega \gg k_\parallel v_{\text{th}}$, in which case $k_\parallel$ can be omitted. In this case the density response may be written

$$\frac{\delta n}{n} = -\frac{q\phi}{T}\left[1 - \frac{1}{n_0}\int_0^\infty \frac{\omega - \omega_*[1 + \eta(mv^2/2T - 3/2)]}{\omega - \omega_D(v_\parallel^2 + v_\perp^2/2)/v_{\text{th}}^2} J_0(\xi) f_0 \, \mathrm{d}^3v\right] \quad (5.129)$$

where ω_D is the fluid magnetic drift velocity and all velocity dependence has been written explicitly. For comparison with fluid theory it is useful to expand equation (5.129) for $\omega_D/\omega \ll 1$ and $\xi = kv_{\text{th}}/\Omega_c \ll 1$. Including terms up to second order in both small parameters we have

$$\frac{\delta n_i}{n} = \left[\frac{\omega_{*e}}{\omega} - \left(1 - \frac{\omega_{*iT}}{\omega}\right)\left(1 + \frac{7}{4}\frac{\omega_{Di}}{\omega}\right)\left(\frac{\omega_{De}}{\omega} + k^2\rho^2\right) + \frac{7}{4}\eta_i \frac{\omega_{*i}\omega_{De}\omega_{Di}}{\omega^3}\right]\frac{e\phi}{T_e} \quad (5.130)$$

where $\omega_{iT} = \omega_{*i}(1 + \eta_i)$.

We note that equation (5.130) is also useful for MHD modes, since for these ions can usually be treated in the electrostatic limit. For these modes the natural linear eigenfrequency is ω_{*iT}, at which the second part of equation (5.130) vanishes. The last term here acts as an additional driving pressure force, which is responsible for an instability below the MHD beta limit [5.70]. As can be seen from the advanced fluid model presented later, the last term in equation (5.130) is due to the divergence of the diamagnetic heat flow, which is the term in the energy equation that corresponds to the lowest order driving term in the continuity equation, i.e., the divergence of the diamagnetic particle flux. Another property of equation (5.130) is that to first order in ω_*/ω and ω_D/ω it reduces to $\omega_*(1 - \varepsilon_n)/\omega$, where $\varepsilon_n = \omega_D/\omega_*$. Since ω_{*e}/ω leads to the main driving term for interchange and ballooning modes in MHD, the ε_n part is the main reason for the reduction of the growth rate of MHD ballooning modes for large ε_n seen in kinetic theory [5.70].

5.8 Trapped Particle Instabilities

In a tokamak, the magnetic field consists of a toroidal (along the torus) and a poloidal (around the cross-section) component. Thus, the magnetic field lines are wound in the way shown in figure 5.7.

When the small radius a is much smaller than the large radius R ($a \ll R$), the magnetic surface, defined as the surface described by the field line during

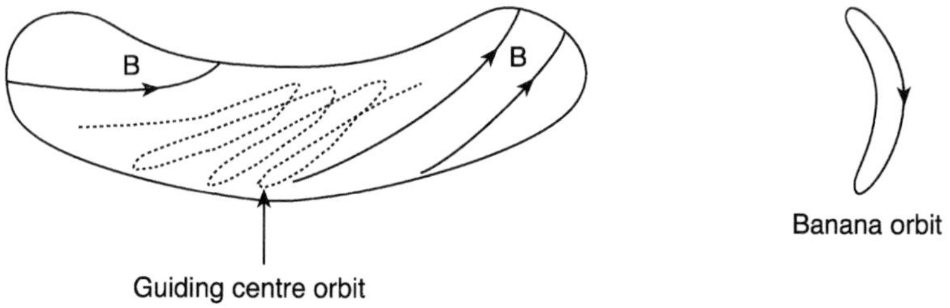

Figure 5.7. Trapped particle orbits.

many turns around the torus, has an almost circular cross-section. The toroidal magnetic field, however, is decreasing in the direction from the centre (along R). This means that a magnetic mirror is formed. A particle with small enough velocity along B will then be trapped in such a way that it never reaches the point $\theta = \pi$. In fact, it turns out that particles with $v_\parallel < \sqrt{2\varepsilon}v_\perp$, where $\varepsilon = r/R$, become trapped. Assuming an isotropic velocity distribution function we then conclude that a fraction $\sqrt{2\varepsilon}$ of the particles will be trapped. Clearly, such trapping effects may decrease the possibilities for electrons to cancel space charge by moving along the magnetic field. Another important effect of the trapping is, however, to increase the effective collision frequency. Normally, the collision frequency ν corresponds to 90 degrees scattering. For trapped particles, however, a scattering angle of $\sqrt{\varepsilon}$ is clearly very significant since it may lead to detrapping. This may be accounted for by introducing a collision frequency $\nu_{\text{eff}} = \nu/\varepsilon$.

When studying systems with trapped particles we have to treat trapped and untrapped particles separately. For the study of trapped electrons we use the drift kinetic equation (5.105) in the electrostatic approximation. We shall, however, include a magnetic mirror force as can be obtained by including equation (5.106) in equation (5.102). Assuming an anisotropic Maxwellian distribution with $T_\perp \gg T_\parallel$ (relevant for trapped particles), we can see that the $\partial f/\partial v_\perp$ part can be neglected. We then obtain, including a Krook collision term

$$\frac{\partial f_{Te}}{\partial t} + \mathrm{i}k_\parallel v_\parallel f_{Te} - \mathrm{i}\frac{1}{B_0}k_y\phi\frac{\partial f_{0Te}}{\partial x} + \left(\frac{e}{m}\mathrm{i}k_\parallel\phi - \frac{\mu}{m}\frac{\partial B_0}{\partial z}\right)\frac{\partial f_{0Te}}{\partial v_\parallel}$$
$$= -\nu_{\text{eff}}\left(f_{Te} - \frac{e\phi}{T}f_{0Te}\right). \tag{5.131}$$

The collision term relaxes the distribution function to a Maxwellian at potential ϕ in a time ν_{eff}^{-1}. In the force along B_0 we have included the effect of the inhomogeneity of B. This force is proportional to the magnetic moment μ. In order to have a strong influence of the trapping we realize that we must assume $\omega_B \gg \omega$, where ω_B is the bounce frequency of particles due to trapping.

If this condition is fulfilled the trapping may prevent the thermalization of the particles in the wave field. The particles then see a stationary field during a bounce period. Since for a closed orbit a contribution $v_\parallel f_{Te}\,\mathrm{d}t$ to the orbit integral will then be cancelled by an equal contribution, where $v_\parallel \to -v_\parallel$, we realize that the orbit average

$$\int_0^{2\pi/\omega_B} v_\parallel f_{Te}\,\mathrm{d}t = 0.$$

Introducing

$$\frac{\partial f_{0Te}}{\partial v} = -\frac{m_e}{T_e} v_\parallel f_{0Te}$$

we find that the orbit average of the fourth term is also zero. This follows from the fact that the energy exchange $\int F v_\parallel\,\mathrm{d}t$ is zero where F is the force in equation (5.131). We then arrive at the averaged equation

$$\frac{\partial f_{Te}}{\partial t} - \mathrm{i}\frac{1}{B_0}k_y\phi\frac{\partial f_{0Te}}{\partial x} = -\nu_{\mathrm{eff}}\left(f_{Te} - \frac{e\phi}{T} f_{0Te}\right). \tag{5.132}$$

Since we have now removed all explicit $v_\parallel$ dependence we can integrate equation (5.132) over $v_\parallel$, thus replacing f_{Te} by δn_e and f_{0Te} by n_{0Te}. Now introducing

$$\frac{\partial f_{0Te}}{\partial x} = -\kappa f_{0Te}$$

and

$$\omega_{*e} = \frac{k_y \kappa T_e}{e B_0}$$

we obtain (with $\partial/\partial t = -\mathrm{i}\omega$)

$$-\mathrm{i}(\omega + \mathrm{i}\nu_{\mathrm{eff}}) f_{Te} + \mathrm{i}(\omega_{*e} + \mathrm{i}\nu_{\mathrm{eff}})\frac{e\phi}{T} f_{0Te} = 0.$$

Then, considering the relation integrated over $v_\parallel$, we have the trapped electron perturbation

$$\frac{\delta n_{Te}}{n_{0Te}} = \frac{\omega_{*e} + \mathrm{i}\nu_{\mathrm{eff}}}{\omega + \mathrm{i}\nu_{\mathrm{eff}}}\frac{e\phi}{T}. \tag{5.133}$$

Assuming that the untrapped electrons thermalize (reach a Boltzmann distribution), we arrive at the electron density

$$\frac{\delta n_e}{n_0} = \left[\sqrt{\varepsilon}\frac{\omega_{*e} + \mathrm{i}\nu_{\mathrm{eff}}}{\omega + \mathrm{i}\nu_{\mathrm{eff}}} + (1 - \sqrt{\varepsilon})\right]\frac{e\phi}{T_e} \tag{5.134}$$

where $\sqrt{\varepsilon}$ is the fraction of trapped electrons.

If the bounce frequency of the ions, ω_{Bi}, fulfils $\omega_{Bi} \ll \omega$, we may disregard the effect of trapping on the ions. The ion density is then

$$\frac{\delta n_i}{n_0} = \left(\frac{\omega_{*e}}{\omega} - k_y^2 \rho^2 + \frac{k_\parallel^2 c_s^2}{\omega^2} \right) \frac{e\phi}{T_e}. \tag{5.135}$$

Now, using the quasi-neutrality condition and treating ν_{eff}, $k_y^2 \rho^2$ and $k_\parallel^2 c_s^2/\omega^2$ (but not $\sqrt{\varepsilon}$) as small, we arrive at the dispersion relation

$$\omega \approx \omega_{*e} \left(1 - \frac{k_y^2 \rho^2 - k_\parallel^2 c_s^2/\omega_{*e}^2}{1 - \sqrt{\varepsilon}} \right) + \mathrm{i} \frac{\nu_{\text{eff}} \sqrt{\varepsilon}}{\omega_{*e}(1 - \sqrt{\varepsilon})} (\omega_{*e} - \omega). \tag{5.136}$$

Assuming the solution $\omega = \omega_r + \gamma$, where $\gamma \ll \omega_r$, we now find the growth rate

$$\gamma = \frac{\nu_{\text{eff}} \sqrt{\varepsilon}}{\omega_{*e}(1 - \sqrt{\varepsilon})^2} \left(k_y^2 \rho^2 - \frac{k_\parallel^2 c_s^2}{\omega_{*e}^2} \right). \tag{5.137}$$

We thus find that the growth rate is modified by the factor $\sqrt{\varepsilon}(1 - \sqrt{\varepsilon})^{-2}$, in addition to the effect of trapping on the effective collision frequency $\nu_{\text{eff}} = \nu/\varepsilon$. This instability is the trapped electron instability. When $\omega_{Bi} > \omega$, we may also have a trapped ion instability. Because the trapped particle distribution behaves as if $k_\parallel = 0$, i.e., as for a flute mode, the trapped particle modes may also be driven unstable by a magnetic curvature.

In the presence of an electron temperature gradient a new branch of this mode is introduced by trapping. This mode is believed to be responsible for the Alcator scaling of the energy confinement time in a tokamak.

5.9 Reactive Drift Modes

The eigenmodes that we have considered until now have basically either been of the drift type, characterized by nearly Boltzmann distributed electrons, or of the MHD type, characterized by small or zero parallel electric field. As shown in chapter 3, the MHD modes are, in general, of a more global character and often show reactive instability, i.e., instability without dissipative effects. The drift modes, on the other hand, in general require dissipation to become unstable. The reason for this is that Boltzmann-distributed electrons are free to move along field lines to cancel space charge. Accordingly, the charge separation caused by gravity or magnetic drifts is cancelled and the interchange instability does not occur.

There exists, however, a third class of modes between the MHD modes and the usual drift waves. This class may be called reactive drift modes, and typically has the maximum growth rate for $k^2 \rho^2 \sim 0.1$. Since the ideal MHD modes generally have to be stable in fusion machines, the reactive drift modes, which are the second most dangerous class of modes, are potentially the most

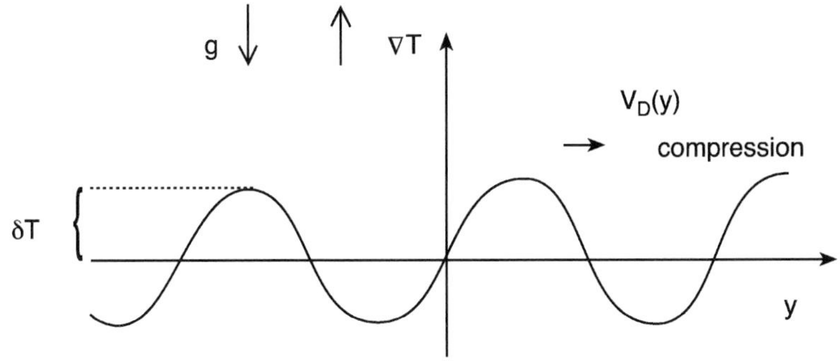

Figure 5.8. Compression due to a gravitational drift.

likely candidates for explaining the observed transport in tokamaks. The first derivation of this new class of modes was made by Rudakov and Sagdeev [5.1], when they discovered the slab η_i mode in equation (5.152). This was, in fact, also the first work on drift waves as a whole. Later, the trapped particle modes, which also belong to this class, were discovered by Kadomtsev and Pogutse [5.10].

5.9.1 Ion temperature gradient modes

There are basically two ways in which a reactive instability can be recovered for drift waves. The first has already been indicated in the previous section. If some of the electrons are trapped they will not be able to cancel space charge and an interchange instability is recovered. As a second possibility, we notice that when a real curvature is used, an interchange mode is driven by the full pressure gradient (see equation (5.48)). Here the temperature gradient part does not correspond to a charge separation, but rather a compressibility. In the fluid sense, a compressibility comes about as a divergence of a velocity. A velocity with a divergence has to vary in its own direction, thereby causing local rarefractions and bunchings. Since the convective part of $\nabla \cdot (n\boldsymbol{v}_*)$ is cancelled by a part of $\nabla \cdot \boldsymbol{v}_*$, the full driving pressure term appears as a compressibility. If, however, we replace $\boldsymbol{v}_*$ by a gravity drift, where the temperature is perturbed, it becomes clear that it is the temperature perturbation part which is associated with compressibility:

$$\nabla \cdot (n\boldsymbol{v}_g) = \boldsymbol{v}_g \cdot \nabla n + n\nabla \cdot \boldsymbol{v}_g$$

where now

$$\nabla \cdot \boldsymbol{v}_g = \frac{1}{T}\boldsymbol{v}_g \cdot \nabla \delta T.$$

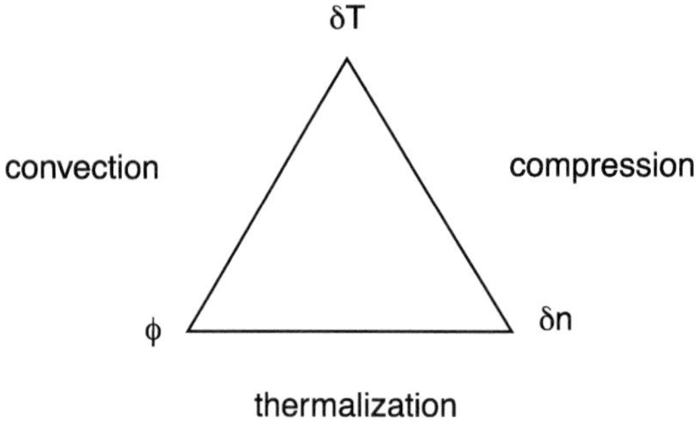

Figure 5.9. Feedback loop of thermal instability.

When δT is due to $E \times B$ convection in a background temperature gradient, i.e.

$$\delta T = -\frac{\omega_*}{\omega}\eta q\phi \tag{5.138}$$

where

$$\eta = \frac{\mathrm{d}\ln T}{\mathrm{d}\ln n} = \frac{L_n}{L_T} \qquad L_n = -\left(\frac{1}{n}\frac{\mathrm{d}n}{\mathrm{d}r}\right)^{-1} \qquad L_T = -\left(\frac{1}{T}\frac{\mathrm{d}T}{\mathrm{d}r}\right)^{-1}$$

we obtain the dynamics shown in figure 5.8. Here, as usual, the x and y directions correspond to the r and θ directions in a torus. The variation of v_g along its direction gives rise to a density perturbation. We now assume Boltzmann electrons (3.3)

$$\frac{\delta n_e}{n} = \frac{e\phi}{T_e} \tag{5.139}$$

while the ions are subject to the compressibility. Using quasi-neutrality we then obtain a feedback mechanism, as shown in figure 5.9.

Whether there is a positive or negative feedback depends on the relative directions of g and ∇T. Not surprisingly, it turns out that the feedback is positive (destabilizing) when g and ∇T have opposite directions, i.e., in unfavourable curvature regions.

Now combining equations (5.23) and (5.138) for the ions we obtain

$$\nabla \cdot (n\boldsymbol{v}_{*i}) = \frac{1}{T_i}\boldsymbol{v}_{Di} \cdot \nabla\left[T_i\delta n - \frac{\omega_{*i}}{\omega}\eta_i en\phi\right]. \tag{5.140}$$

Also using equation (5.24), we now obtain from the ion continuity equation

$$\frac{\delta n_i}{n} = \frac{\omega_{*e} + \tau\omega_{Di} - \tau\eta_i(\omega_{Di}\omega_{*i}/\omega)}{\omega - \omega_{Di}}\frac{e\phi}{T_e}. \tag{5.141}$$

When combined with equation (5.139) this gives the dispersion relation

$$\omega[\omega - \omega_{*e} - \omega_{Di}(1+\tau)] = -\tau\eta_i\omega_{*i}\omega_{Di}. \tag{5.142}$$

In the unstable case, equation (5.142) may be written $\omega = \omega_r + \mathrm{i}\gamma$, where

$$\omega_r = \frac{1}{2}\left[\omega_{*e} - \omega_{De}\left(1+\frac{1}{\tau}\right)\right] \tag{5.143}$$

$$\gamma = \omega_{*e}\varepsilon_n^{1/2}\sqrt{\eta_i - \eta_{i\mathrm{th}}} \tag{5.144}$$

where $\varepsilon_n = \omega_D/\omega_*$, and

$$\eta_{i\mathrm{th}} = \frac{1}{4}\left[1 - \varepsilon_n\left(1+\frac{1}{\tau}\right)\right]. \tag{5.145}$$

The present exercise merely serves to show that there is a reactive instability for large η_i. The magnetic drift terms have not been treated consistently here and several others should enter equation (5.138), as will be shown later. The threshold (5.145) is thus incorrect. In early treatments the denominator in (5.141) was expanded for $\omega_D/\omega \ll 1$, and the ω_D/ω term combined with ω_{*e} in the numerator. This leads to

$$\frac{\delta n_i}{n} = \frac{1}{\omega}\left[\omega_{*e} - \tau(1+\eta_i)\frac{\omega_{Di}\omega_{*i}}{\omega}\right]\frac{e\phi}{T} \tag{5.146}$$

where the $\tau\omega_{Di}$ term is also neglected as compared to ω_{*e}. This corresponds to using the convective density response directly in equation (5.140). If the stabilizing linear term in ω is now ignored, equation (5.146) leads to the stability threshold $\eta_i = -1$, which is often quoted in the literature. In this case, part of the ω^2 term necessary for an instability has been obtained artificially by an expansion in ω_D/ω. This introduces a spurious instability for $\eta_i = 0$. The correct threshold is usually around $\eta_i = 1$, as will be shown later. The instability obtained here is a reactive drift instability driven by the temperature gradient and magnetic curvature. The mode is usually referred to as the toroidal η_i mode [5.28, 5.61, 5.74].

We have shown how an instability is obtained when the compressibility originates from the divergence of v_{gi}. The original η_i mode instability was, however, obtained as a result of the compressibility associated with the parallel ion motion. The feedback scheme in figure 5.9 also applies in this case.

For the parallel ion motion we take

$$\frac{\partial v_\parallel}{\partial t} = -\frac{e}{m_i}\frac{\partial\phi}{\partial z} - \frac{1}{m_i n}\frac{\partial P_i}{\partial z} \tag{5.147}$$

leading to

$$v_{\parallel i} = \frac{k_\parallel}{\omega m_i}\left(e\phi + \delta T_i + T_i\frac{\delta n}{n}\right). \tag{5.148}$$

Now, again using equation (5.138) for δT_i we obtain

$$v_{\parallel i} = \frac{k_\parallel}{\omega m_i}\left[\left(1 - \eta_i \frac{\omega_{*i}}{\omega}\right) e\phi + T_i \frac{\delta n}{n}\right]. \tag{5.149}$$

Including parallel ion motion in our derivation of the toroidal η_i mode we then obtain

$$\frac{\delta n_i}{n} = \left[\omega_{*e} + \tau\omega_{Di} - \tau \frac{\omega_{*i}\omega_{Di}}{\omega}\eta_i + \frac{k_\parallel^2 c_s^2}{\omega}\left(1 - \eta_i \frac{\omega_{*i}}{\omega}\right)\right]\left(\omega - \omega_{Di} - \frac{k_\parallel^2 c_s^2}{\tau\omega}\right)^{-1}. \tag{5.150}$$

In order to consider the excitation due to parallel ion motion separately, we now put $\omega_D = 0$. The dispersion relation may then be written

$$\omega^3 - \omega^2\omega_{*e} - \omega k_\parallel^2 c_s^2 \left(1 + \frac{1}{\tau}\right) + \eta_i \omega_{*i} k_\parallel^2 c_s^2 = 0. \tag{5.151}$$

The driving term here is the last term, and the simplest possible dispersion relation giving the instability is

$$\omega^3 = -\eta_i \omega_{*i} k_\parallel^2 c_s^2. \tag{5.152}$$

Since $\omega_{*i} < 0$, ω^3 is positive for positive η_i. Taking the phase angle as 2π we obtain an unstable root with phase angle $2\pi/3$. This instability, which does not require curvature, is usually referred to as the slab instability (slab mode) since its eigenvalue can be treated in slab geometry [5.1].

The η_i mode is among the most serious candidates for explaining the anomalous ion heat transport in present day tokamaks. This may be anticipated already by its fundamental nature as a thermal instability. When we heat a glass of water from below, we generate convection through a thermal instability. When we heat a tokamak with a centrifugal force due to field curvature, we have a corresponding situation and a similar thermal instability may develop. The toroidal version has the largest growth rate in the bulk of the plasma, while the slab version may have larger growth rate close to the edge, where $\omega_D \ll \omega_*$. The slab version usually has a slightly lower threshold, while the parallel ion motion is stabilizing when the toroidal drive dominates [5.74, 5.92, 5.166]. Fully kinetic treatments show that both modes have their maximum growth rate around $k_\perp^2 \rho_i^2 = 0.1$.

5.9.2 Electron temperature gradient mode

A mode that is sometimes used to try to explain the anomalous electron and heat transport in the collisionless regime is the electron temperature gradient mode (η_e mode). This mode also exists in both slab [5.21] and toroidal [5.97, 5.115] versions. It is a very short wavelength mode fulfilling

$$\rho_e \ll \lambda \ll \rho_i.$$

In this limit the ions are unmagnetized and, furthermore, in the hot regime. We may thus take

$$\frac{\delta n_i}{n} = -\frac{e\phi}{T_i}. \tag{5.153}$$

The ions reach thermal equilibrium by moving perpendicular to the magnetic field. This also requires $\omega < k_\perp v_{\mathrm{th}\,i}$, which can now also be fulfilled with $\omega > \Omega_{ci}$. The large mode number makes the frequency comparatively large. All that we shall require here is that

$$\omega \ll \Omega_{ce}. \tag{5.154}$$

In this regime we may still use the drift expansion for the electrons, but it may be possible to ignore parallel electron motion. This means that as compared to the η_i mode the roles of ions and electrons are switched. We may thus follow the previous procedure. The electron density response (corresponding to equation (5.141) is then

$$\frac{\delta n_e}{n} = \frac{\omega_{*e} - \omega_{De} + \eta_e(\omega_{De}\omega_{*e}/\omega)}{\omega - \omega_{De}}\frac{e\phi}{T_e}. \tag{5.155}$$

In combination with equation (5.153) we now obtain the dispersion relation

$$\omega(\omega + \omega_{*e} - 2\omega_{De}) = -\eta_e\omega_{De}\omega_{*e}. \tag{5.156}$$

This dispersion relation is very similar to equation (5.142). An important difference is that the η_i mode propagates in the electron drift direction and the η_e mode propagates in the ion drift direction for small $|\omega_D|$. A correct treatment of the ω_D terms shows that for realistic values of $\varepsilon_n = \omega_D/\omega_*$, the η_i mode propagates in the ion drift direction and the η_e mode in the electron drift direction. Such a trend can also be seen in our present treatment, which is, however, not accurate enough to justify such a conclusion.

Due to the very short wavelength, the η_e mode only gives a small direct transport. It can, however, excite modes with longer wavelength through mode coupling. Such modes with a wavelength of the order of the skin depth c/ω_{pe} can give a neo-Alcator scaling (cf section 5.1). The slab version of this mode is analogous to that of the η_i mode although we shall not discuss it here.

5.9.3 Trapped electron modes

The most obvious candidates for explaining the large anomalous electron thermal conductivity in tokamaks are the trapped electron modes [5.10, 5.14]. As mentioned previously, trapped electron modes can give a neo-Alcator scaling in the collision dominated regime. In the collisionless regime an interchange type of mode driven by the density gradient is often referred to as the ubiquitous mode. We shall here consider a collisionless trapped electron mode, which is similar to the η_e mode but occurs for $k^2\rho^2 \sim 0.1$. Taking the fraction of the trapped electrons to be f_t, we may use the response (5.155) for the trapped electrons since their motion along the magnetic field has been averaged out. We then have to consider the magnetic drift to be bounce averaged. The free electrons are assumed to be Boltzmann distributed. Then

$$\frac{\delta n_e}{n} = f_t \frac{\omega_{*e} - \omega_{De} + \eta_e(\omega_{De}\omega_{*e}/\omega)}{\omega - \omega_{De}} \frac{e\phi}{T_e} + (1 - f_t)\frac{e\phi}{T_e}. \tag{5.157}$$

Now using equation (5.141) for the ion response we obtain

$$\frac{\omega_{*e} - \omega_{De} + \eta_i(\omega_{De}\omega_{*e}/\omega) - k^2\rho_s^2(\omega - \omega_{*iT})}{\omega - \omega_{Di}}$$
$$= f_t \frac{\omega_{*e} - \omega_{De} + \eta_e(\omega_{De}\omega_{*e}/\omega)}{\omega - \omega_{De}} + 1 - f_t. \tag{5.158}$$

This relation shows a symmetry between ions and trapped electrons. We note that equation (5.158) is now a cubic equation in ω. This means that it has at least one real root and, accordingly, a maximum of two complex conjugate roots, i.e., it can have no more than one unstable mode. The more complex fluid description in section 5.11 gives a quadratic equation and, accordingly, the possibility of having two unstable modes at the same time.

For that system it is possible to consider resonant modes where $\omega \sim \omega_{De}$, and in that way one may decouple the ion density perturbation. Here we shall denote the left-hand side of equation (5.158) by Δ. The dispersion relation for the trapped electron mode may then formally be written

$$\omega\left(\omega + \frac{f_t}{\xi}\omega_{*e} - \frac{1-\Delta}{\xi}\omega_{De}\right) = -\frac{f_t}{\xi}\eta_e\omega_{De}\omega_{*e} \tag{5.159}$$

where $\xi = 1 - f_t - \Delta$.

Equation (5.159) is very similar to the dispersion relation (5.156) for the η_e mode. The dispersion relation shows only the electron dynamics and is a relevant description when the ion dynamics are subdominant. The ω^2 term is here entirely due to electron dynamics and we may have an instability driven only by electron compressibility and temperature gradient. This dispersion relation is accordingly analogous to equation (5.142) for the η_i mode, which is destabilized by ion dynamics alone. In fact, if we take the limit $\Delta \to 0$ in equation (5.159), the two modes are symmetric for $\tau = 1$ except for the factor f_t appearing in equation (5.159).

In order to investigate the other modes given by equation (5.158), we rewrite it in its cubic form

$$\begin{aligned}
&\omega(\omega - \omega_{De})(\omega - \omega_{*e} + \omega_{De} - \omega_{Di}) \\
&\quad - \omega\left[\omega_{Di}\omega_{*e}\frac{\eta_i + \tau f_t \eta_e}{1 - f_t} - \frac{k^2\rho_s^2}{1 - f_t}(\omega - \omega_{De})(\omega - \omega_{*iT})\right. \\
&\quad \left. + \frac{f_t}{1 - f_t}(\omega_{Di} - \omega_{De})(\omega_{*e} - \omega_{De})\right] \\
&= \frac{\eta_e f_t - \eta_i}{1 - f_t}\omega_{Di}\omega_{De}\omega_{*e}.
\end{aligned} \tag{5.160}$$

Here the single ω factor on the left-hand side is associated with the temperature gradients and, indeed, if the temperature gradients vanish, so does the right-hand side, ω factors out and we obtain a quadratic dispersion relation. On the other hand, we also note that the right-hand side is quadratic in the magnetic drifts. We may thus neglect it for this reason, thus obtaining a quadratic dispersion relation. Then, also neglecting other terms that are quadratic in the magnetic drift we obtain the dispersion relation

$$\begin{aligned}
&(\omega - \omega_{De})\left[\omega\left(1 + \frac{k^2\rho^2}{1 - f_t}\right) - \omega_{*e} - \omega_{iT}\frac{k^2\rho^2}{1 - f_t} + \omega_{De} - \omega_{Di}\right] \\
&= \frac{f_t}{1 - f_t}\omega_{*e}(\omega_{Di} - \omega_{De}) + \frac{\eta_i + \tau f_t \eta_e}{1 - f_t}\omega_{*e}\omega_{Di}.
\end{aligned} \tag{5.161}$$

We note that in the limit $f_t \to 0$, equation (5.155) is similar to equation (5.136) although the ω^2 term has a different origin. The differences are the inclusion of the FLR effect and the Doppler shift $\omega - \omega_{De}$ in the first ω factor. The latter difference is due to the nonadiabatic electron response, which was absent in the derivation of equation (5.142). It is important to note that in the absence of both trapping and temperature gradients there is no instability, i.e., the product of frequency-independent parts on the left-hand side of equation (5.161) cannot drive an instability. (Compare the discussion after equation (5.146).) Thus, equation (5.161) is most conveniently solved by introducing $\bar{\omega} = \omega - \omega_{De}$ and first obtaining the solution for ω. In the absence of temperature gradients or compressibility, equation (5.161) gives a pure trapped electron mode. This mode, which may propagate either in the electron or ion drift direction, depending on the values of $k^2\rho_s^2$ and ε_n, is usually called the *ubiquitous mode* [5.19, 5.28]. The ω^2 term there requires nonadiabatic responses from both ions and electrons. The ubiquitous mode is, in fact, stabilized by temperature gradients, as we shall see in section 5.11. If we multiply equation (5.161) by $(1 - f_t)$ and take the limit $f_t \to 1$ we obtain a pure MHD equation in the limit $k_\parallel = 0$.

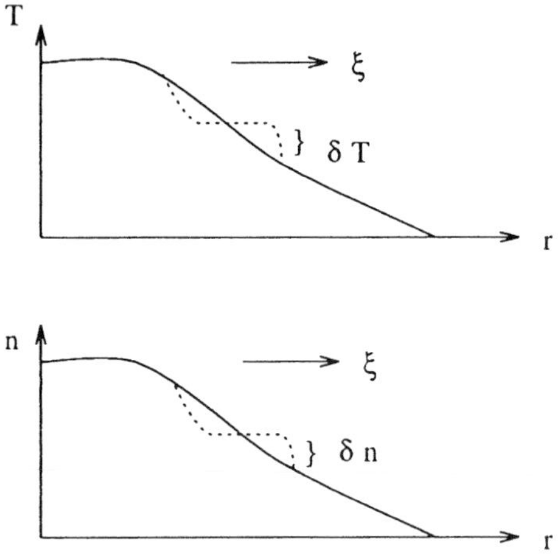

Figure 5.10. Convective perturbations of density and temperature.

5.9.4 Competition between inhomogeneities in density and temperature

As we have seen in equations (5.142) and (5.151), also very simple models for the temperature gradient mode also indicate that the parameter $\eta = L_n/L_T$ is critical for stability. This is partly because the diamagnetic drift $1/L_n$ introduces a real eigenfrequency that is stabilizing. A more fundamental reason for the importance of the parameter η is, however, the simultaneous convection in temperature and density gradients. This leads to a competition between convection and expansion (negative compression) in the energy equation. When the convection is outward, higher density parts move out into more dilute areas where the expansion takes place. The expansion cools the plasma and competes with the increase in temperature due to temperature convection

$$\delta T = -\boldsymbol{\xi} \cdot \nabla T + \alpha \boldsymbol{\xi} \cdot \nabla n.$$

Here α is a coefficient that gives the expansion cooling. From this relation we immediately see that η has to exceed a certain limit for δT to be positive. A corresponding equation for n is obtained because a temperature perturbation, through v_D or $v_\parallel$, leads to compression. Thus

$$\delta n = -\boldsymbol{\xi} \cdot \nabla n + \beta \boldsymbol{\xi} \cdot \nabla T$$

where we considered the convective temperature perturbation. The competition between temperature and density gradients, in the nonlinear regime, leads to inward contributions to fluxes. When several sources of free energy are present (coupled relaxations), we may even have net inward fluxes (pinches) of some thermodynamic variables. The total energy flux is, however, always directed outward. A realistic threshold, including the effects discussed here, will be derived in the next section.

5.10 Advanced Fluid Models

One of the main problems with creating a first principles transport model, which can be used in transport codes, is the fact that because of the resonance

$$\omega = k_{\parallel} v_{\parallel} + \omega_D(v_{\parallel}^2, v_{\perp}^2) \tag{5.162}$$

kinetic theory is, in principle, needed. On the other hand, not even the most efficient computers are able to run a fully nonlinear kinetic code as part of a transport code. In fact, nonlinear kinetic simulations are usually made on time scales of the order of linear growth time γ^{-1} and nonlinear saturation, which is typically a few growth times, while transport codes operate on time scales of the order of the confinement time τ_E. While the growth time is typically of the order of 10^{-5} s, τ_E is of the order of seconds. Thus, what we are left with for transport simulations is either kinetic models that ignore velocity space nonlinearities, or some kind of advanced fluid model that attempts to incorporate the resonance (5.162) in some approximate way. The latter possibility has only been explored in the last decade.

5.10.1 The development of research

The beginning of the development of advanced fluid theories, of course, depends on how we define the concept. With our definition, as will be given shortly, it dates back to 1986, with the first published papers appearing in 1987.

Before this time, all fluid models expanded the dynamic equations such that $\omega_D/\omega \ll 1$ (adiabatic state) for the perpendicular dynamics, and introduced an equation of state with a free parameter γ that can describe adiabatic or isothermal states for the parallel motion. As it turns out, however, when kinetic or advanced fluid theories are used, ω and ω_D are usually comparable, except at the edge. Because of this, all previous drift-wave theories had a basic flaw in that transport coefficients decreased with radius in the models, while they increased with radius in the experiments. As an example, we may mention the work by Scott *et al* [5.110], where the radial profiles of ion thermal conductivity from two one-pole fluid models were compared with the experimental ion thermal conductivity for a TFTR shot.

Although, in general, the parallel part of the resonance (5.162) may be important, we shall focus here first on the perpendicular part, which is associated with the very fundamental toroidal effects. As a result of toroidicity, the eigenmodes tend to localize on the outside of the torus, where the curvature is unfavourable. The ultimate limit, which, when it can be reached, gives the largest growth rate, is the local limit, where the mode is strongly localized and effects of the parallel dynamics vanish. In this limit, the eigenvalue equation turns into an algebraic dispersion relation.

Comparisons between local kinetic theory and a two-pole fluid model [5.108] showed that the diamagnetic heat flow q_*, as given by equation (2.27), through the magnetic inhomogeneity part of its divergence, reproduces the main kinetic effects of ω_D in equation (5.162). By including this term in the fluid equations we obtain a two-pole fluid density response in the local limit and recover both adiabatic and isothermal limits for the perpendicular dynamics. This was done in the advanced fluid model developed at Chalmers University of Technology in 1986, as described in section 5.11. It was first developed from the local limit of an electromagnetic model [5.90, 5.91], but later the electrostatic eigenvalue problem was also solved [5.92]. The ion thermal conductivity, based on quasi-linear theory and mode-coupling simulations, was published in 1988 [5.93, 5.94]. The increased order of the fluid response due to $\nabla \cdot \boldsymbol{q}_*$ is significant, since it changes the regions of positive and negative energy modes. This can be seen from the expression

$$\varepsilon(\omega, k) = \frac{1}{k^2\lambda_{de}^2}\left(\frac{\delta n_i}{n_i} - \frac{\delta n_e}{n_e}\right) \tag{5.163}$$

for the dielectric function in combination with the expression for the wave energy

$$W = \frac{1}{4}\frac{\partial}{\partial\omega}(\omega\varepsilon)|\nabla\phi|^2. \tag{5.164}$$

As an example, we note that the electromagnetic version of this fluid model reproduces the instability of the MHD ballooning mode branch below the ideal MHD beta limit in the presence of an ion temperature gradient [5.90]. This instability is due to $\nabla \cdot \boldsymbol{q}_*$ and is caused by a shift of regions of negative and positive energy. From the drift wave point of view, $\nabla \cdot \boldsymbol{q}_*$ introduces a new stability regime, with positive wave energy for large ε_n. This has the effect of giving a strong trend for χ_i to grow towards the edge, as shown by figure 5.11. The new regime where ε_n is stabilizing is generally termed 'the flat density regime'. Since ε_n decreases towards the edge, the system departs more from marginal stability as we move towards the edge if the density and temperature profiles have similar shapes. *The flat density regime typically prevails in the inner 80% of tokamak discharges, which means that the new regime is dominant and radically changes the predictions of drift wave theory*. The TFTR shot studied in [5.110] was also studied by the advanced fluid model described in the next section, giving a χ_i which followed the experimental trend over the whole cross-section. An upper stability regime in ε_n can also be obtained due to $\nabla \cdot \boldsymbol{v}_E$ in an adiabatic model [5.98]. This stabilization is, however, of an FLR type, similar to that discussed in section 4.1.4, since the wave energy is still negative. As will be shown in section 5.11.6, parallel ion motion destabilizes

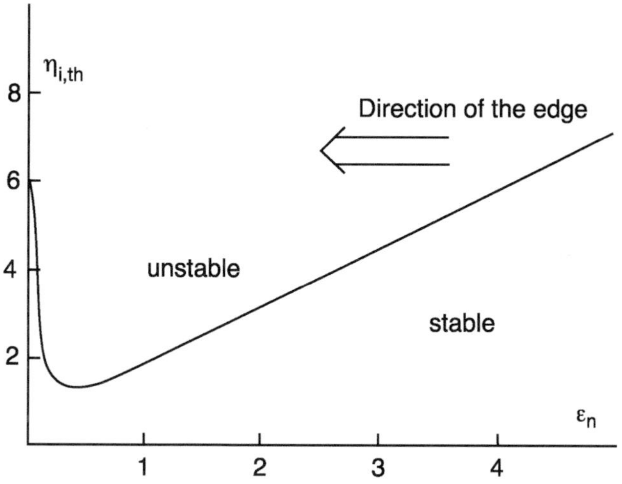

Figure 5.11. The stability diagram in ε_n and η_i showing the stability regime for large ε_n, and the trend for increasing deviation from marginal stability as we move towards the edge for similar shapes of density and temperature profiles.

this regime by introducing a dissipative damping due to magnetic shear. We shall thus introduce the following definition of an advanced fluid model:

An advanced fluid model is a fluid model which recovers the stable regime of η_i modes for large ε_n due to positive wave energy.

In H modes the density profile is flat over a large part of the discharge. This regime was investigated in [5.89], and it was pointed out that the critical parameter for stability here is R/L_{Ti} rather than η_i. As it turns out, most of the cross-section of L modes is also usually in the regime where this type of stability criterion applies. Such a regime was present in the general stability criterion of [5.91], but was not pointed out until [5.93] and evaluated as $L_T/R = 0.367$. In [5.98] the FLR type stabilization in the adiabatic limit was obtained with the threshold $L_T/R = 0.28$, and in [5.101] the full local kinetic result $L_T/R = 0.35$ was obtained. It is interesting to note that a local kinetic model using a gradient B approximation of the magnetic drift [5.102] gives the threshold $L_T/R = 0.375$, which has a larger deviation from the exact kinetic result than the advanced fluid model in [5.91]. The threshold $L_T/R = 0.350$ was also obtained independently in [5.108]. We moreover note that the quasi-linear correspondence to the upper stability regime in ε_n is a pinch flux proportional to ε_n, as seen in equation (5.185). Here only the $\nabla \cdot \boldsymbol{q}_*$ part contributes, i.e., the part that changes the sign of the wave energy. The stabilizing effects of $\boldsymbol{q}_*$ are stronger in the hot ion regime. This is true both linearly and for the pinch flux. This is, in fact, the main reason for the good confinement in the hot ion regime in this type of theory. We finally note that as a consequence of the upper stability regime in ε_n, the toroidal η_i mode is stable near the axis in tokamaks. This was first pointed out in [5.96].

5.10.2 Closure

The reason for the truncation of the above advanced fluid model, by taking $\boldsymbol{q} = \boldsymbol{q}_*$, is that $\boldsymbol{q}_*$ is the highest moment that depends only on the moments that are normally fed by sources (fuelling, heating) in a magnetic confinement device. In general we have

$$\langle v_i v_j v_k v_l \rangle = \langle v_i v_j \rangle \langle v_k v_l \rangle + \cdots + G(\boldsymbol{r}, t) \tag{5.165}$$

where $G = \langle v_i v_j v_k v_l \rangle_{\text{irr}}$ is the irreducible part. The transport equation for G can be written in the form

$$\frac{\partial G}{\partial t} = \frac{1}{r}\frac{\partial}{\partial r}\left(r\chi_G \frac{\partial G}{\partial r} \right) + S_G. \tag{5.166}$$

The formal procedure in deriving the fluid model is to approximate the four-velocity correlation in the heat flow equation with products of two velocity correlations [5.137], which means taking $G = 0$. Higher order moments (i.e., G above) will not have sources in their transport equations ($S_G = 0$), and should decay on a timescale of the order of the confinement time, while the moments that are fed by sources remain in quasi-stationary states for many confinement times. Thus, taking $G = 0$ leads to one diamagnetic heat flow for parallel temperature and one for perpendicular temperature, given by equations (5.27) and (5.28). For isotropic temperature the Braginskii $\boldsymbol{q}_*$ is recovered as the sum of these. As it turns out, the energy equations for parallel and perpendicular temperatures contain nonlinearities that tend to isotropize the temperature perturbations.

A phenomenon which will be discussed in the following is that of inward 'pinch' fluxes. Since these have a tendency to equilibrate length scales (such as $L_T = -T/(\partial T/\partial r)$), we note that a higher moment at a low level becomes very sensitive (as both T and $\partial T/\partial r$ are small), and can easily adjust itself to an equilibration of length scales without affecting the lower moments. An eventual pinch in the transport equation for the higher moment would thus not replace a source.

A relation that holds for temperatures is that the perturbation becomes small if the background is small. Thus, extrapolating this relation to the irreducible part of the perturbed four-velocity correlation (δG), we expect it to decay to zero in a confinement time. With this closure, which on timescales longer than the confinement time, according to the above arguments, will be valid, we can treat the whole range of states from adiabatic to isothermal, i.e., with arbitrary relations between frequency and magnetic drift frequency.

In the local limit, the ion density response is now a two-pole response and when parallel ion motion is included it becomes a three-pole response. When we include higher fluid moments, the order of the density response increases by one for each new moment. The fluid resonances become more and more densely packed as we increase the order until they form a continuum in the infinite limit. The product of infinitely many fluid resonances in the denominator leads to a kinetic, dissipative resonance

$$\frac{\delta n}{n} = \frac{\omega - \omega_* + \cdots}{(\omega - \alpha_1 \omega_D)(\omega - \alpha_2 \omega_D) \cdots (\omega - k_\parallel v_{\mathrm{th}}) \cdots} \frac{e\phi}{T_e}. \tag{5.167}$$

The fluid truncation used here thus includes the fluid resonances that correspond to moments with sources in the experiment. These resonances form a part of the kinetic resonance. We thus include the part of the kinetic resonance that corresponds to the moments that are maintained by external sources. This part is then treated self-consistently in the transport calculations.

Another related aspect of the fluid hierarchy is that higher order moments are much more sensitive to lower moments than vice versa. One example of this is the well known feature that heat flows are much more sensitive to the temperature profiles than temperature profiles are to the heat flows. Experience from dealing with higher order linear moments in the local limit (e.g. from [5.137]) shows that the introduction of a new, higher order moment leads to a large shift in the dispersion function when the former eigenvalue is used, but a small shift in the eigenvalue is sufficient to restore the dispersion function to its previous value. Thus, the higher order moment is very sensitive to the eigenvalue. This may be due to the fact that new poles are introduced by the higher order moment. Comparisons with kinetic nonlocal theory [5.121, 5.166] show that higher order moments have a greater impact on the eigenvalue

We also note that the complication of an integral eigenvalue problem in kinetic theory [5.140, 5.148] is absent in the fluid theory. The only approximation in the fluid theory is associated with the truncation. In the nonlocal theory with parallel ion motion, it turns out that the difference in linear threshold between the kinetic theory and the reactive fluid model can be rather large [5.121, 5.166] when s/q is of order 1. This means that the properties of the fluid model depend on how this discrepancy, mainly due to linear Landau damping, is treated. Our advanced fluid model just ignores this difference in linear theory and only retains moments that can be treated self-consistently. The closure made thus relies on the decay of the moment G on the transport time scale. With this nonlinear closure, the fluid moments kept do not have to converge towards a smaller influence for higher moments. On the contrary, the highest moment kept, $\boldsymbol{q}_*$, is one of the most important parts. Some aspects of the velocity space dynamics with potential importance to the closure are discussed in section 5.10.4. Since the closure described here does not make use of dissipation we shall call this type of fluid model a 'reactive fluid model'.

5.10.3 Gyro-Landau fluid models

Gyro-Landau fluid models are a class of fluid models that take a radically different point of view on the closure problem from that presented above. This class of models is actually somewhat beyond the main scope of the present review and we shall only make a brief survey here without claims of completeness.

Development of gyro-Landau fluid models was initiated by the work of Hammet and Perkins on Landau damping in the fluid equations for the slab η_i mode [5.116]. This work introduces Landau damping through an imaginary parallel heat flow q in the energy equation, and is able to recover linear kinetic results for the slab η_i mode. A follow-up paper discussed the details of the closure and how the result depends on the level in the fluid hierarchy at which the dissipation is introduced [5.117]. Toroidal effects, and with them, magnetic drift resonances were introduced by Waltz, Dominguez and Hammet [5.142]. This work also included FLR effects to all orders. The fluid equations were derived by taking moments of the gyrokinetic equation (5.127) and agreement was obtained with the reactive fluid model described in the previous section in the appropriate limit. The closure in this gyro-Landau model can be written

$$q = q_* + \mathrm{i}\boldsymbol{q}_{gl} \tag{5.168}$$

where $\boldsymbol{q}_{gl}$ represents the contribution to the resonance from infinitely many higher order moments, as obtained by a fit to a Maxwellian velocity distribution. This model also gives very good agreement with linear kinetic theory. Turbulence simulations in three dimensions have also been performed with this model [5.155].

The fundamental assumption in gyro-Landau models is that the gyro-Landau resonance, obtained by a fit to linear kinetic theory, can be used in transport models operating in a nonlinearly saturated state. More recent gyro-Landau fluid models [5.157, 5.163] make the closure at a higher level in the fluid hierarchy, but the basic principle of closure is the same.

5.10.4 Nonlinear kinetic fluid equations

A more complete approach, which can be seen as intermediate to the usual gyro-Landau models and kinetic theory, is the analytical solution to the Vlasov equation obtained by Mattor and Parker [5.164] in slab geometry. Here the closure is nonlinear although the background velocity distribution function still is Maxwellian. Resonant particles are assumed to follow the phase velocity of the waves so that an integration over particle velocities can be replaced by an integration over wave phases. This model preserves time reversibility and can support a type of trapping oscillation, where the velocity distribution is fixed but the wave phase velocity oscillates due to a periodic nonlinear frequency shift. It leads to a considerably lower time-averaged saturation level than the Hammet–Perkins theory, and to time reversible oscillations after the nonlinear saturation. The maxima of these trapping oscillations are close to the Hammet–Perkins saturation level, so the reason for the difference is, in fact, that the Hammet–Perkins model phase locks at the maxima of the trapping oscillations, while the average level in the Mattor–Parker model corresponds to an averaging over the trapping oscillations. Such oscillations can also be expected to occur when higher order moments relax to a nonlinear equilibrium state. If we include inertia of the resonant particles so that they do not follow the wave phases exactly, we would expect additional phase mixing and relaxation to a stationary state. This state would be the attractor, where the average force between resonant particles and waves changes sign. In systems with many waves, we would expect much stronger phase mixing of resonant particle orbits and more quasi-linear behaviour.

5.10.5 Comparisons with nonlinear gyrokinetics

Comparisons between gyro-Landau models and nonlinear gyrokinetics have been going on for several years [5.154]. Recently, the Cyclone group in the US compared both the magnitude of χ_i and the stiffness (how rapidly χ_i increases with the temperature gradient above threshold) of several models. The trend for gyro-Landau models to give too large a transport (up to a factor of 3 above the gyrokinetic level) is similar to the results found in [5.164], where the analytic solution by Mattor and Parker was very close to the full kinetic saturation level, while the Hammet–Perkins saturation level was far above. The global and flux-tube gyrokinetic simulations gave somewhat different results in that the flux tube simulations gave more transport and larger stiffness. One of the main questions that has been discussed regarding the saturation level is the damping due to nonlinearly generated background flows [5.181]. Since these flows have a stronger effect on longer wavelengths, the presence of longer wavelengths in the global simulations may create a more absorbing boundary condition for these, as compared to the situation in the flux tube simulations. We note that in this respect the reactive fluid transport model, as described in section 5.11.3, should rather be compared to the global gyrokinetic simulations since an absorbing boundary condition for long wavelengths was used. It is also likely that the mixing length leading to the type (3.67) diffusion coefficient is essential for the stiffness. The scaling $\chi \sim \eta_i - \eta_{i\,\mathrm{th}}$, just above threshold, has been seen in mode coupling simulations [5.94, 5.100, 5.105], and was also derived analytically in [5.130] in the flat density regime. An important point is that in the comparison by Mattor and Parker, the same three-wave system was used for all models so the boundary conditions in k-space were also the same. This leaves only the different closure schemes as a reason for the differences. On the other hand, since this system contained only three modes, it corresponded to a much more coherent situation than that in the gyrokinetic simulations, where a broad spectrum of modes was included. In the Cyclone simulations, some comparisons with the reactive fluid model discussed above have also been made. Preliminary results were reported in [5.183].

5.11 Reactive Fluid Model for Strong Curvature

As mentioned several times above, a more complete fluid model is needed in order to obtain a correct threshold for η_i, η_e and trapped electron modes. In typical fusion plasmas the magnetic drifts are also comparable to the diamagnetic drifts, except close to the edge. As pointed out above, this is the reason for the development of advanced fluid models. A circumstance that improves the possibility for fluid models is that the magnetic drift causes a stream in the plasma. Because of this, a fluid resonance, similar to the fluid-beam plasma resonance, is present. Another favourable aspect is that magnetic drifts do not appear explicitly in fluid equations unless the temperature is anisotropic, in which case the curvature drift appears. Without magnetic drifts it is clear that the magnetic field localizes the particles in the perpendicular direction and that the parameter $k^2\rho^2$ can be chosen as a small parameter to truncate the fluid hierarchy. The usual truncation of the fluid hierarchy is that by Braginskii. It assumes collision dominance so that the perturbation of the velocity distribution function is Maxwell distributed. In combination with the expansion in $k^2\rho_i^2$, this leads to the so-called Righi–Leduc or diamagnetic heat flow

$$q = q_* = \frac{5}{2}\frac{P}{m\Omega_c}(\hat{e}_\parallel \times \nabla T) \tag{5.169}$$

to lowest order in $k^2\rho_i^2$. In this fluid model the temperature is isotropic. Recently, a collisionless fluid model was derived by truncating the irreducible part of the fourth moment in the heat-flow equation [5.137]. In this model no assumptions of Maxwellian distribution were made and temperatures were just defined through quadratic velocity moments. This model gave different $\boldsymbol{q}$ for transport of parallel and perpendicular energy in equations (5.27) and (5.28) but when the parallel and perpendicular temperatures were assumed equal the Braginskii energy equation with the heat flow $\boldsymbol{q}_*$ given by equation (5.169) was recovered. Although the temperatures are anisotropic in collisionless linear theory, the isotropic fluid model gives good agreement with Vlasov theory for the toroidal η_i mode concerning threshold, and rather good agreement concerning growth rate in the local limit [5.108]. The main reason for this seems to be that in the low beta case the average of the parallel and perpendicular temperatures enters the driving pressure term. When parallel ion motion is important, however, anisotropy is essential. We here rely on the nonlinear closure discussed above.

5.11.1 The toroidal η_i mode

The energy equation is written

$$\frac{3}{2}n_i\left(\frac{\partial}{\partial t} + \boldsymbol{v}_i \cdot \nabla\right)T_i + P_i\nabla \cdot \boldsymbol{v}_i = -\nabla \cdot \boldsymbol{q}_{*i}. \tag{5.170}$$

Now

$$\nabla \cdot \boldsymbol{q}_{*i} = -\tfrac{5}{2}n\boldsymbol{v}_{*i} \cdot \nabla T_i + \tfrac{5}{2}n\boldsymbol{v}_{Di} \cdot \nabla T_i. \tag{5.171}$$

Here the first convective diamagnetic part cancels with other convective diamagnetic terms after substitution of the continuity equation for $\nabla \cdot \boldsymbol{v}_i$, as shown in chapter 2. We shall here retain the curvature part of $\nabla \boldsymbol{q}_{*i}$, which will turn out to be very important. The linearized temperature perturbation is now

$$\frac{\delta T_i}{T_i} = \frac{\omega}{\omega - 5\omega_{Di}/3}\left[\frac{2}{3}\frac{\delta n_i}{n} + \frac{\omega_{*e}}{\omega}\left(\eta_i - \frac{2}{3}\right)\frac{e\phi}{T_e}\right]. \tag{5.172}$$

Using equation (5.172) instead of equation (5.138), equation (5.141) is replaced by

$$\frac{\delta n_i}{n_i} = [\omega(\omega_{*e} - \omega_{De}) + (\eta_i - 7/3 + 5\varepsilon_n/3)\omega_{*e}\omega_{Di} - k^2\rho_s^2(\omega - \omega_{*iT})(\omega - 5\omega_{Di}/3)][\omega^2 - 10\omega\omega_{Di}/3 + 5\omega_{Di}^2/3]^{-1}\frac{e\phi}{T_e} \tag{5.173}$$

where

$$\omega_{*iT} = \omega_{*i}(1 + \eta_i)$$

and we also included the polarization drift and the lowest order FLR effect, as derived in chapter 2. The response (5.173) is of higher degree in ω_D than equation (5.141), both in numerator and denominator.

The most important improvement in equation (5.173) is that it has the correct asymptotic limit for large ω_D, i.e., in the isothermal limit

$$\frac{\delta n_i}{n} \to -\frac{e\phi}{T_i} \tag{5.174}$$

for $\omega_D \gg \omega, \omega_*$. This can be obtained from equation (5.141) only in the absence of a temperature gradient. Since δT influences equation (5.141) only through the temperature gradient, one can conclude that a careful treatment of the energy equation is required to make the fluid theory consistent with kinetic theory in the presence of temperature perturbations. The key property of equation (5.172), absent in equation (5.138), is that we obtain the correct isothermal limit $\delta T_i \to 0$ when $\omega_D \gg \omega, \omega_*$. This is entirely due to the curvature part of $\nabla \cdot \boldsymbol{q}_*$. This part enters as an additional higher order contribution to the pressure force that may be either destabilizing or stabilizing. The response (5.173) was first applied to MHD ballooning modes [5.90], where $\nabla \cdot \boldsymbol{q}_*$ reproduced an instability below the MHD beta limit, previously only seen in kinetic treatments [5.70]. For ion temperature gradient modes it is usually stabilizing [5.91]. It is instructive to compare the expansion of equation (5.173) in ω_D/ω with the corresponding expansion of the gyrokinetic equation (5.130). These expansions are identical except for the replacement of 7/4 by 5/3. It is fortunate that the terms of order $k^2\rho^2\omega_D/\omega$ agree, since no attempt was made to systematically include these in equation (5.173). The last term proportional to η_i is asymmetric with respect to $\omega - \omega_{*iT}$. It represents a correction of the basic MHD pressure balance and is, accordingly, responsible for the instability below the MHD beta limit seen in kinetic theory. The fluid model obtained here has sometimes been called fully toroidal since it does not expand in ε_n. Effects of ε_n are, in fact, the most important toroidal effects on drift waves. We emphasize here that the truncation in equation (5.169) is treated as exact in the present fluid model. This means that we assume equations (5.172) and (5.173) to be valid for arbitrary ω/ω_{Di}, and these equations should not, in general, be expanded. Clearly, the fact that we keep the frequency dependence in equation (5.172) means that we can describe both slow and fast processes. This increases the number of non-zero poles in the density response of equation (5.173) by one.

By using equation (5.173) in combination with equation (5.139) we obtain the dispersion relation

$$\omega = \omega_r + \mathrm{i}\gamma \tag{5.175}$$

where

$$\omega_r = \frac{1}{2}\omega_{*e}\left[1 - \varepsilon_n\left(1 + \frac{10}{3\tau}\right) - k^2\rho_s^2\left(1 + \frac{1+\eta_i}{\tau} - \varepsilon_n - \frac{5}{3\tau}\varepsilon_n\right)\right] \tag{5.176}$$

and

$$\gamma = \frac{\omega_{*e}\sqrt{\varepsilon_n/\tau}}{1 + k^2\rho_s^2}\sqrt{\eta_i - \eta_{i\text{th}}} \tag{5.177}$$

where

$$\begin{aligned}\eta_{i\text{th}} = {} & \frac{2}{3} - \frac{\tau}{2} + \varepsilon_n\left(\frac{\tau}{4} + \frac{10}{9\tau}\right) + \frac{\tau}{4\varepsilon_n} \\ & - \frac{k^2\rho_s^2}{2\varepsilon_n}\left[\frac{5}{3} - \frac{\tau}{4} + \frac{\tau}{4\varepsilon_n} - \left(\frac{10}{3} + \frac{\tau}{4} - \frac{10}{9\tau}\right)\varepsilon_n + \left(\frac{5}{3} + \frac{\tau}{4} - \frac{10}{9\tau}\right)\varepsilon_n^2\right].\end{aligned} \tag{5.178}$$

Here, equation (5.178) was expanded in $k^2\rho^2$. For very small ε_n, terms of the type $k^4\rho^4/\varepsilon_n$ and with higher powers in ε_n have not been calculated consistently, since $\boldsymbol{q}_*$ has been obtained only to lowest order in $k^2\rho_s^2$. They do, however, give the correct trend when compared with kinetic theory. The marginal stability curve in an η_i, ε_n diagram is shown in figure 5.10. An important feature, as compared to the previous fluid threshold, is the presence of an upper stability regime in ε_n. For large ε_n, $\tau = 1$ and $k^2\rho_s^2 \to 0$ this threshold is

$$\eta_{i\text{th}} = 1.36\varepsilon_n = 2.72L_n/L_B. \tag{5.179}$$

Here L_n scales out and the stability condition becomes

$$L_{Ti} > 0.367L_B. \tag{5.180}$$

The correct kinetic threshold here is $0.35L_B$. The large ε_n regime is often the relevant regime in the bulk of tokamak discharges. This is so in particular for H mode discharges, which usually have flat density profiles. The upper stability regime in ε_n also determines how close to the axis the η_i mode can be unstable. Another interesting aspect of equation (5.176) is the presence of an $k^4\rho_i^4\eta_i^2$ term in ω_r^2, which enters the stability criterion to the next order in $k^2\rho^2$. Since this term does not contain ε_n, it is in fact consistent and gives an upper stability regime in η_i. This has led to enhanced confinement states in transport code simulations [5.114]. In conclusion, we thus find that toroidal effects introduce a completely new regime for large ε_n which is dominant in the bulk of tokamaks. The neglect of this regime caused discrepancies between η_i mode theory and experiments for a long time. The new philosophy in the present fluid model, as compared to simple fluid models, is that the truncation is made by equation (5.172) and is assumed to be valid for both slow and fast processes.

5.11.2 Electron trapping

Since the kinetic integrals for trapped electrons and ions without parallel motion are symmetric [5.9], we may use the same fluid model for the trapped electrons as for the ions. Introducing the fraction of trapped electrons f_t, we obtain the dispersion relation [5.103, 5.109]:

$$\begin{aligned}
&\frac{\omega_{*e}}{N_i}\left[\omega(1-\varepsilon_n)+\left(\eta_i-\frac{7}{3}+\frac{5}{3}\varepsilon_n\right)\omega_{Di}\right.\\
&\left.\quad - k^2\rho_s^2[\omega-\omega_{*i}(1+\eta_i)]\left(\frac{\omega}{\omega_{*e}}+\frac{5}{3\tau}\varepsilon_n\right)\right]\\
&= f_t\frac{\omega_{*e}}{N_e}\left[\omega(1-\varepsilon_n)+\left(\eta_e-\frac{7}{3}+\frac{5}{3}\varepsilon_n\right)\omega_{De}\right]+1-f_t \qquad (5.181)
\end{aligned}$$

where

$$N_j=\omega^2-\tfrac{10}{3}\omega\omega_{Dj}+\tfrac{5}{3}\omega_{Dj}^2, \quad j=i,e. \qquad (5.182)$$

Here the denominator N_j acts as the resonant denominator in the dispersion relation of a two-stream instability. When $N_i < N_e$, the mode propagates in the ion direction (η_i mode), and when $N_e < N_i$, the mode propagates in the electron direction (trapped electron mode). Equation (5.181) is the generalization of equation (5.158) to arbitrary ε_n, and is a quadratic equation. Accordingly, it can have two modes unstable at the same time. For ε_n of order 1 the modes are rather independent, propagating in opposite directions, and the dispersion relations can usually be well approximated by neglecting the part with the larger N_j in equation (5.181). For small ε_n, however, the modes are strongly coupled and the directions of propagation may change. For large ε_n and $\eta_i \sim \eta_e$ the η_i mode is the most unstable of the modes. Then,

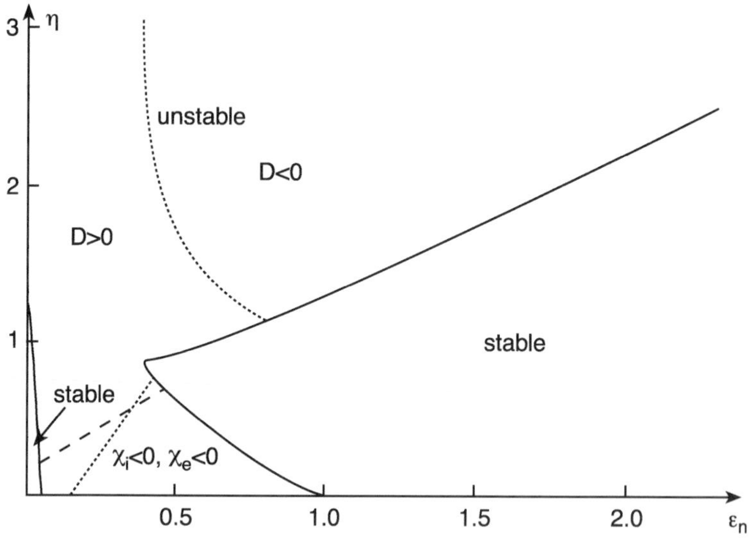

Figure 5.12. Stability diagram in ε_n and η ($\eta_i = \eta_e = \eta$). (From Weiland J, Jarmén A and Nordman H 1989 *Nucl. Fusion* **29** 1810, courtesy of the IAEA.)

ignoring the trapped electron part with denominator N_e we obtain the stability threshold

$$L_{Ti} > L_B \left[\frac{20}{9\tau}(1 - f_t) + \frac{\tau}{2(1 - f_t)} \right]^{-1} \tag{5.183}$$

which is the generalization of equation (5.180) for finite electron trapping. If we instead take N_i large we obtain a generalization of equation (5.159) where $\Delta = 0$. This is actually the only way of isolating a trapped electron mode which can be driven only by compressibility and the electron temperature gradient. This was first done in [5.103]. Since this mode is obtained for $N_e \ll N_i$, it is due to a fluid resonance. A corresponding mode due to the kinetic resonance was discussed by Adam *et al* [5.23]. In the same sense, the toroidal η_i mode may also be regarded as resonant.

It is clear from this discussion that these modes require a description valid for $\omega \sim \omega_D$. The stability boundaries for $\eta_i = \eta_e = \eta$ are shown in figure 5.12. The modes present in this system are most clearly shown (figure 5.13) if we display the growth rates as a function of η for an ε_n where the modes are separated in figure 5.12. We can see the 'ubiquitous' (trapped electron) mode [5.19, 5.28] for small η. The toroidal η_i mode becomes unstable at η just above 1, and we then have the compressional trapped electron mode that becomes unstable at η just above 2.

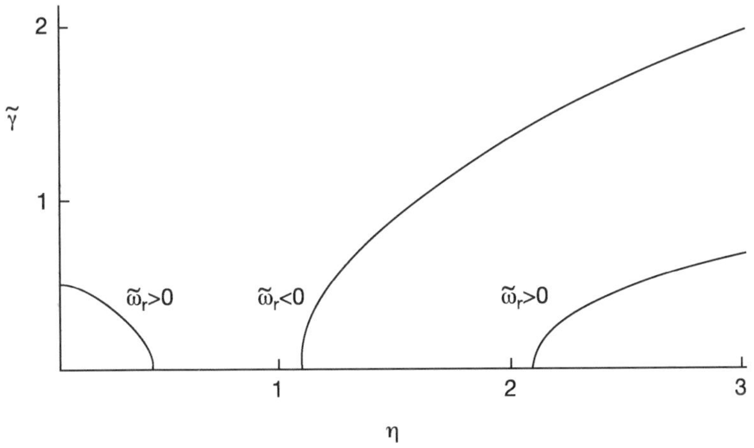

Figure 5.13. Growth rate as a function of η for $\varepsilon_n = 0.8$ ($\tilde{\omega} = \omega/\omega_{*e}$). Other parameters are the same as in figure 5.12. (From Nilsson J and Weiland J 1995 *Nucl. Fusion* **35** 497, courtesy of the IAEA.)

The compressional trapped electron mode tends to dominate the transport when $\eta_e > \eta_i$, which is typical when most heating goes to electrons. This is the case for alpha particle heating in burning plasmas.

5.11.3 Transport

We may calculate the quasi-linear ion thermal conductivity in the same way as we calculated the particle diffusion in chapter 2. The thermal conductivity is calculated using Fick's law from

$$\Gamma_T = -\chi \frac{\mathrm{d}T}{\mathrm{d}x} \tag{5.184}$$

and the saturation level in equation (3.65) can also be derived from the energy equation. The result for the toroidal η_i mode without parallel ion motion is [5.93, 5.100]

$$\chi_i = \frac{1}{\eta_i}\left(\eta_i - \frac{2}{3} - \frac{10}{9\tau}\varepsilon_n\right)\frac{\gamma^3/k_x^2}{(\omega_r - 5\omega_{Di}/3)^2 + \gamma^2}. \tag{5.185}$$

Equation (5.185) has been shown to give good agreement with mode-coupling simulations using many modes if the fastest growing mode ($k^2\rho^2 = 0.1$) is used [5.94, 5.100]. The pinch terms (with negative sign) in equation (5.185) are important. In particular, the ε_n pinch term (due to $\nabla \cdot \boldsymbol{q}_*$) significantly improves the agreement between experimental and theoretical radial profiles of

χ_i by suppressing χ_i in the inner region where ε_n is large. The temperature diffusion obtained from equation (5.185) is always outward, since γ would be zero if the pinch effects dominated. If we also use the complete fluid model for the trapped electrons, we obtain, as mentioned above, a quadratic dispersion relation, where both the η_i mode and the trapped electron mode are included and may be unstable simultaneously [5.103, 5.109]. In this system there is a coupling between the diffusion of T_i, T_e and n, with a trend to equilibrate the equilibrium length scales L_{Ti}, L_{Te} and L_n. This system contains a possibility of inward fluxes (pinch effects) but the total pressure flux is always outward [5.109]. It is also interesting to note that in the edge of tokamaks, typically $\varepsilon_n \ll 1, \eta_i > 1$ and $\gamma > \omega_r$. In this limit, equation (5.185) gives the well known mixing length expression $\chi = \gamma/k_x^2$. For the full system with electron trapping the agreement with experimental tokamak transport is remarkably good. In particular, the radial profiles of both χ_e and χ_i are usually in rough agreement with experiments, at least for $r/a < 0.8$, where a is the small radius, and the magnitude is also usually of the right order.

Furthermore, the ratio χ_e/χ_i is typically about 1/2 at half radius with a radial growth that is somewhat faster than that of f_t. This is also a very typical experimental situation. For the TFTR supershots with peaked density profile and τ about 0.3, χ_e/χ_i is usually small, often less than 1/4, and for the D-III-D hot ion mode with $\tau \sim 0.2$ and flat density profile, large values of χ_e/χ_i are obtained (typically about 4). Both these cases are well reproduced by the fluid model with electron trapping. In self-consistent transport code simulations, this model also gives the L mode scaling of the energy confinement time with heating power in rough agreement with equation (1.8) and, moreover, a spontaneous transition to an H mode (enhanced confinement regime) with an improvement of τ_E by a factor between 2.5 and 3 for sufficiently strong heating [5.114].

The transport coefficients for T_i, T_e and n_e with electron trapping included can be written as

$$\chi_i = \frac{1}{\eta_i}\left[\eta_i - \frac{2}{3} - (1-f_t)\frac{10}{9\tau}\varepsilon_n - \frac{2}{3}f_t\Delta_i\right]\frac{\gamma^3/k_x^2}{(\omega_r - 5\omega_{Di}/3)^2 + \gamma^2} \tag{5.186}$$

$$\chi_e = f_t\frac{1}{\eta_e}\left(\eta_e - \frac{2}{3} - \frac{2}{3}\Delta_e\right)\frac{\gamma^3/k_x^2}{(\omega_r - 5\omega_{De}/3)^2 + \gamma^2} \tag{5.187}$$

$$D = f_t\Delta_n\frac{\gamma^3/k_x^2}{\omega_{*e}^2} \tag{5.188}$$

where, introducing $\hat{\omega} = \omega/\omega_{*e}$

$$\Delta_i = \frac{1}{N}\left\{|\hat{\omega}|^2\left[|\hat{\omega}|^2(\varepsilon_n - 1) + \hat{\omega}_r\varepsilon_n\left(\frac{14}{3} - 2\eta_e - \frac{10}{3}\varepsilon_n\right)\right.\right.$$
$$\left. + \frac{5}{3}\varepsilon_n^2\left(-\frac{11}{3} + 2\eta_e + \frac{7}{3}\varepsilon_n\right) - \frac{5}{3\tau}\varepsilon_n^2\left(1 + \eta_e - \frac{5}{3}\varepsilon_n\right)\right]$$

$$
+ \frac{50}{9\tau}\hat{\omega}_r\varepsilon_n^3(1-\varepsilon_n) - \frac{25}{9\tau}\varepsilon_n^4\left(\frac{7}{3}-\eta_e-\frac{5}{3}\varepsilon_n\right)\Bigg\} \tag{5.189}
$$

$$
\Delta_e = \frac{1}{N}\Bigg\{|\hat{\omega}|^2\Bigg[|\hat{\omega}|^2(\varepsilon_n-1)+\hat{\omega}_r\varepsilon_n\left(\frac{14}{3}-2\eta_e-\frac{10}{3}\varepsilon_n\right)
+ \frac{5}{3}\varepsilon_n^2\left(-\frac{8}{3}+3\eta_e+\frac{2}{3}\varepsilon_n\right)\Bigg]
+ \frac{50}{9}\hat{\omega}_r\varepsilon_n^3(\varepsilon_n-1)+\frac{25}{9}\varepsilon_n^4\left(\frac{7}{3}-\eta_e-\frac{5}{3}\varepsilon_n\right)\Bigg\} \tag{5.190}
$$

$$
\Delta_n = \frac{1}{N}\Bigg[|\hat{\omega}|^2(1-\varepsilon_n)-\hat{\omega}_r\varepsilon_n\left(\frac{14}{3}-2\eta_e-\frac{10}{3}\varepsilon_n\right)
- \frac{5}{3}\varepsilon_n^2\left(-\frac{11}{3}+2\eta_e+\frac{7}{3}\varepsilon_n\right)\Bigg] \tag{5.191}
$$

$$
N = \left(\hat{\omega}_r^2-\hat{\gamma}^2-\frac{10}{3}\hat{\omega}_r\varepsilon_n+\frac{5}{3}\varepsilon_n^2\right)^2+4\hat{\gamma}^2\left(\hat{\omega}_r-\frac{5}{3}\varepsilon_n\right)^2. \tag{5.192}
$$

The signs of these in various parameter regimes are indicated in figure 5.11.

The most remarkable result obtained with the transport coefficients in equations (5.186)–(5.188) is that from the simulation [5.129] of the heat pinch on D-III-D [5.128]. In this simulation, the ECH electron heat source was at half radius and ions were only heated by collisions with electrons. The ECH and ohmic heating of electrons were taken from the experiment, while the particle source at the edge was taken as a free parameter. Both the electron energy pinch and the density and temperature profiles were well reproduced in this simulation. The electron energy pinch was driven by the η_i mode. Other mechanisms, associated with the toroidal curvature and trapped electrons, have also been suggested [5.161].

5.11.4 Normalization of transport coefficients

In deriving the transport coefficients in equations (5.186)–(5.188) we used the saturation level in equation (3.65). This saturation level was obtained by balancing linear growth with nonlinear effects at the correlation length scale [5.93]. Thus the nonlinear effects are here entirely stabilizing. This corresponds to a situation where nonlinear mode coupling carries energy and momentum away from the linearly unstable region in k-space and nothing comes back, i.e., we have absorbing boundaries both at short and long space scales. This saturation level has recently been recovered by a non-Markovian Fokker–Planck theory [5.186]. Both the ion thermal conductivity without trapping of equation (5.185), and the transport coefficients in equations (5.186)–(5.188), have been normalized and tested against nonlinear mode coupling simulations with absorbing boundaries, both at short and long scales [5.94, 5.100, 5.109]. Good agreement was obtained when the FLR parameter $k^2\rho^2$ was about 0.1.

This corresponds to the linearly fastest growing mode. Here k should be interpreted as the inverse correlation length. This can be understood by observing that the correlation length is typically determined by the shortest space scales that are strongly excited and this is often given by the source region. It was also verified in mode coupling simulations that the saturation level in equation (3.67) gave better agreement than the mixing length estimate. The damping for long space scales was, in the mode coupling code, obtained from viscocity, while that for long wavelengths was artificial. We note, however, that damping due to sheared background flows increases with the space scale, and thus has the properties we need for absorbing boundaries at large space scales.

5.11.5 Finite Larmor radius stabilization

As seen from equation (5.178), FLR effects are usually rather marginal for the pure η_i mode. When electron trapping is included it does, however, become stronger. A particular limit in which we can see this explicitly is for $\varepsilon_n \ll 1$, $\eta_i, \eta_e \gg 1$, so that $\eta \sim 1/\varepsilon_n$. In this limit the dispersion relation splits into two second degree equations, one for $\omega \sim \omega_*$ and one for $\omega \sim \omega_d$. The first case leads to the dispersion relation [5.114].

$$\Omega^2 - \Omega\left[1 + \frac{10}{3}\varepsilon_n\left(1 - \frac{1}{\tau}\right) - \frac{k^2\rho^2}{\tau}\frac{\eta_i}{\Lambda}\right] = -\frac{\varepsilon_n}{\Lambda}\left(\frac{\eta_i}{\tau} + f_t\eta_e\right) \tag{5.193}$$

where $\Lambda = 1 - f_t + k^2\rho^2$ and $\Omega = \omega/\omega_{*e}$. The growth rate is

$$\Omega_I = \sqrt{\frac{\varepsilon_n}{\Lambda}\left(\frac{\eta_i}{\tau} + f_t\eta_e\right) - \frac{1}{4}\frac{k^4\rho_s^4}{\tau^2}\frac{\eta_i^2}{\Lambda^2}}. \tag{5.194}$$

We note that the FLR stabilization is fourth order in $k\rho_s$ and corresponds to an upper stability regime in η_i. It is actually a stability regime for steep temperature gradients since L_n can be taken out of equation (5.194). For the pure η_i mode this regime typically starts at $\eta_i \sim 50$ for $k^2\rho_s^2 \approx 0.1$. We note, however, that if $f_t \to 1$, $\Lambda \to k^2\rho_s^2$ and the FLR stabilization is only second order in $k\rho_s$ (due to the denominator of the first term). For $f_t \approx 0.6$ a stabilization was obtained in predictive transport simulations for $\eta_i = 15$. This leads to an enhanced confinement regime with an improvement of a factor 2.5 in the confinement time [5.114].

In the enhanced confinement state only the mode with $\omega \sim \omega_D$ remains. Its dispersion relation can be written

$$\Omega^2 - \tfrac{10}{3}\varepsilon_n\xi\Omega = -\tfrac{5}{3}\varepsilon_n^2\delta \tag{5.195}$$

where

$$\xi = \frac{\eta_i - f_t\eta_e}{\eta_i + \tau f_t\eta_e} \tag{5.196}$$

$$\delta = \frac{\eta_i + f_t\eta_e/\tau}{\eta_i + \tau f_t\eta_e}. \tag{5.197}$$

We note that L_n cancels out of both ξ and δ, so this mode is a pure magnetic drift mode in a regime where $L_n \ll L_B$. The direction of propagation depends on the sign of ξ and the mode requires $\nabla \cdot \boldsymbol{q}_*$ for instability. This mode always produces a particle pinch, as is easily seen from equation (5.188).

5.11.6 The eigenvalue problem for toroidal drift waves

We shall now briefly consider the eigenvalue problem of toroidal drift modes. We limit our study to the ion temperature gradient driven mode (η_i mode) with Boltzmann electrons. The description of this mode is obtained by combining the response in equation (5.173) with the influence of parallel ion motion, as described by equation (5.148). This leads, for a parallel wavenumber $k_\parallel$, to the response

$$\begin{aligned}\frac{\delta n_i}{n_i} &= \frac{e\phi}{T_e}\left[\omega(\omega_{*e} - \omega_{De}) + EIH\omega_{*e}\omega_{Di}\right.\\ &\quad + \left.\frac{(k_\parallel c_s)^2}{\omega}\left[\omega - \frac{5}{3}\omega_{Di} - \omega_{*i}\left(\eta_i - \frac{2}{3}\right)\right] - FL\right]\\ &\quad \times \left[\omega^2 - \frac{10}{3}\omega\omega_{Di} + \frac{5}{3}\omega_{Di}^2 - \frac{5}{3\tau}(k_\parallel c_s)^2\left(1 - \frac{\omega_{Di}}{\omega}\right)\right]^{-1}\end{aligned} \tag{5.198}$$

where $FL = k^2\rho^2(\omega - \omega_{*iT})(\omega - 5\omega_{Di}/3)$, $\omega_{*iT} = \omega_{*i}(1 + \eta_i)$ and $EIH = \eta_i - 7/3 + 5\varepsilon_n/3$. The parallel wavenumber $k_\parallel$ now becomes an operator, i.e.

$$\mathrm{i}k_\parallel \rightarrow \boldsymbol{e}_\parallel \cdot \nabla = \frac{1}{qR}\frac{\partial}{\partial\theta} \tag{5.199}$$

but with a simple transformation we can avoid operating on $\omega_D(\theta)$ with $k_\parallel$. Within the ballooning mode formulation [5.39, 5.74], we can interpret θ as an extended poloidal angle, where the other operator $(1/qR)\partial/\partial\theta$ includes both poloidal and radial projections on the parallel direction. The eigenvalue equation can be written in the form

$$\frac{\partial^2\phi}{\partial\theta^2} + h\left\{\left[\Omega - 1 + k_\perp^2\rho_s^2\left(\Omega + \frac{1+\eta_i}{\tau}\right)\right]A(\theta) + \varepsilon_n g(\theta)\right\}\phi = 0 \tag{5.200}$$

where

$$h = 4k_\theta^2\rho_s^2\frac{q^2\Omega}{\varepsilon_n^2} \tag{5.201}$$

$$\Omega = \frac{\omega}{\omega_{*e}} \tag{5.202}$$

$$g(\theta) = \cos\theta + s(\theta - \theta_0)\sin\theta \tag{5.203}$$

$$k_\perp^2 = k_\theta^2(1 + s^2\theta^2) \tag{5.204}$$

and

$$A(\theta) = \frac{\Omega + 5\varepsilon_n g(\theta)/3\tau}{F + G\varepsilon_n g(\theta)} \tag{5.205}$$

$$F = \Omega\left(1 + \frac{5}{3\tau}\right) + \frac{1}{\tau}\left(\eta_i - \frac{2}{3}\right)$$

$$G = \frac{5}{3\tau}\left(1 + \frac{1}{\tau}\right).$$

The boundary conditions of equation (5.200) are $\partial\phi/\partial\theta = 0$ at $\theta = 0$ and $\phi \to 0$ when $\theta \to \infty$.

Here θ_0 is a free parameter which can be chosen to maximize the growth rate. Usually its value is zero but we should keep in mind the possibility of other values. This eigenvalue problem, in general, has to be solved numerically. There exist, however, methods of obtaining approximate solutions in most cases of interest. The most important case where we can obtain an exact analytical solution is the strong ballooning limit, where we can take $g(\theta) \approx g(0) = 1$. The applicability of this approximation is wider than we might expect since the geodesic curvature (second part of g) increases when the normal curvature decreases. As it turns out, for $s = 1$, $g(\theta)$ increases slowly with θ up to about $\theta \approx 2.8$. The first zero of $g(\theta)$ for $s = 1$ is just before $\theta = 3$. This picture is changed radically for small and negative shear, where the η_i mode is considerably more stable. In the strong ballooning limit, equation (5.200) is of the form

$$\frac{\partial^2\phi}{\partial\theta^2} + (\xi + \delta s^2\theta^2)\phi = 0. \tag{5.206}$$

Equation (5.206) has solutions of the form

$$\phi \sim \mathrm{e}^{-\alpha\theta^2}$$

where the θ^2 part gives

$$4\alpha^2 + \delta s^2 = 0$$

or

$$\alpha = \pm\tfrac{1}{2}\sqrt{-\delta s^2}. \tag{5.207}$$

The constant part gives

$$\alpha = \xi/2. \tag{5.208}$$

The formal solution is then

$$\xi = -|s|\mathrm{i}\sqrt{\delta}. \tag{5.209}$$

Here we have used the fact that δ contains ω^2, and that a positive imaginary part of ω must give a localized mode. The right-hand side of equation (5.209) corresponds to a convective shear damping, similar to that obtained for usual drift waves in equation (5.38). The right-hand side of equation (5.209) also contains the only effect of the parallel ion dynamics, so the condition $\xi = 0$ gives the local dispersion relation with the solution in equations (5.175)–(5.178). The dispersion relation takes the form

$$(1 + k_\theta^2 \rho_s^2)\Omega^2 - \left[1 - \varepsilon_n\left(1 + \frac{10}{3\tau}\right) - \mathrm{i}\left(1 + \frac{5}{3\tau}\right)\Theta - k_\theta^2 \rho_s^2 \frac{1}{\tau}\left(1 + \eta_i + \frac{5}{3}\varepsilon_n\right)\right]\Omega + \varepsilon_n\left[\Gamma - \frac{5}{3} + \frac{\mathrm{i}\Theta\Gamma}{\varepsilon_n} + \frac{5}{3\tau^2}(1 + \eta_i)k_\theta^2 \rho_s^2\right] \tag{5.210}$$

where

$$\Gamma = \frac{1}{\tau}\left(\eta_i - \frac{2}{3}\right) + \frac{5}{3}\varepsilon_n\left(1 + \frac{1}{\tau}\right) \tag{5.211}$$

$$\Theta = \frac{\varepsilon_n |s|}{2q}\sqrt{A(0)\left(1 + \frac{1 + \eta_i}{\tau\Omega}\right)}. \tag{5.212}$$

Note that $\varepsilon_n |s|/2q = L_n/L_s$.

Although equation (5.200) is rather complicated, the stability threshold takes a very simple form where FLR does not enter [5.166]. It is (note the difference in the definition of ε_n in [5.166]):

$$\eta_{i\mathrm{th}} = \frac{2}{3} + \frac{10}{9\tau}\varepsilon_n. \tag{5.213}$$

The part of the threshold which is linear in ε_n is here entirely due to $\nabla \cdot \boldsymbol{q}_*$. The part $\tau/4\varepsilon_n$ in equation (5.178) is due to the divergence of the $E \times B$ drift and is removed by the parallel ion motion (there are also other contributions from the divergence of the $E \times B$ drift and $\nabla \cdot \boldsymbol{q}_*$, which cancel in equation (5.178). A trend for a higher threshold for the pure toroidal mode was also seen in [5.74]. We also note that for $k^2\rho^2 = 0$, equation (5.178) agrees with equation (5.213) at $\varepsilon_n = 1$. At this point, the boundary curve equation (5.213) is tangent to the stability boundary in the local limit. The influence of parallel ion motion is thus small for the reactive model in the regime where the strong ballooning approximation applies. Equation (5.213) also exactly defines the threshold in η_i for the first factor in equation (5.185) to be positive. We note that equation (5.185) is still valid, since we eliminated the ion density response so that $k_\parallel$ does not enter explicitly. This is also true for the system in equations (5.186)–(5.188), including electron trapping. Due to the Ω dependence of Θ, equation (5.212) is now a rather complicated function of Ω that requires a numerical solution. The order of magnitude of Θ is, however, usually well described by the first factor $0.5\varepsilon_n|s|/q$.

In a recent work [5.166], it was found that if the total pressure perturbation is used in equation (2.47) for the FLR term, equation (5.212) in fact reduces to only the first factor. This model for FLR does, however, not agree with the kinetic expansion in equation (5.130) for terms of the type $\varepsilon_n k^2 \rho_s^2$. We note also that the slab η_i mode is contained in the present formulation. Because of this, the imaginary part is destabilizing for small ε_n, effectively removing the stability regime for small ε_n. Parallel ion motion also has the effect of reducing the threshold slightly in the flat density regime. This is, however, usually a very small effect and the growth rate is somewhat reduced in the region of instability of the local mode. Another interesting result is that the threshold in η_i reduces to 2/3, independent of ε_n in the adiabatic limit, i.e., when we ignore $\boldsymbol{q}_*$. The adiabatic model accordingly does not produce the upper stability regime in ε_n when parallel ion motion is included. Since, as was pointed out above, the linear kinetic threshold (and, in fact, also the threshold of gyro-Landau fluid models) changes considerably when parallel ion motion is included, we find that our reactive model is less sensitive to parallel ion motion than both the simpler adiabatic model and the kinetic model in the linear theory.

5.11.7 Recent tests of the reactive fluid model

The reactive fluid transport model described here has recently been tested against experiments in more complete versions, i.e., including impurities, collisions on trapped electrons, electromagnetic effects, elongation and Shafranov shifts. The most successful overall results have been obtained by the multi-mode model (MMM) in the US [5.160, 5.174, 5.175, 5.182], which includes the reactive fluid model in the good confinement region. The MMM uses an artificial (empirical) dependence on elongation, which is required for a good overall agreement with the database. The pure reactive fluid model has been found to give very good agreement with JET results [5.171, 5.176]. In particular, very good agreement has been obtained with high performance shots when finite beta and elongation effects are included.

The properties of the reactive model can be characterized as: thermodynamic properties (power scalings, stiffness), very good; geometry properties (magnetic q and shear, elongation scalings), fairly good magnetic shear scaling, while the elongation poses problems.

One particular aspect of the thermodynamic properties is that the power scaling has generally been found to be in good agreement with experiment. In particular, the scaling $\tau_E \sim P^{-2/3}$, which is often quoted in the literature, is the worst case obtained with the reactive fluid model. It is obtained for equal ion and electron heating. The exponent −0.5 has been obtained for mainly ion heating, which is in agreement with some experiments using only neutral beam heating, as discussed in [5.114]. The good thermodynamic properties are also emphasized by the agreement with scans in temperature gradient in the Cyclone tests against nonlinear gyrokinetic simulations [5.183]. This was made with the basic electrostatic version for the pure η_i mode. The result can be recovered from equation (5.185) multiplied by 3/2 (energy diffusion), using the local eigenvalue, i.e., from the theory in [5.93]. We note that the background density and temperature profiles were kept fixed in the Cyclone simulations. This is equivalent to applying ideal sources in density and temperature that exactly balance the transport.

The electron trapping effects have also recently been compared with linear kinetic theory [5.187] and tested against perturbative experiments [5.188].

5.12 Resistive Edge Modes

The modes considered so far in chapter 5 have been of a collisionless type, which is the most relevant approximation for the core of tokamak plasmas. At the edge, however, the turbulence changes character and collisionless modes are generally not able to explain the continued growth of the transport coefficients outside 80% of the small radius. In the strongly collisional edge ($\nu_{\text{eff}} \gg \omega$), we note that equation (5.133) predicts Boltzmann distributed trapped electrons. This means that trapping is not important. When ν_{eff} becomes larger than the bounce frequency, the trapped electrons behave as free electrons and the most relevant description is to treat all electrons as free and include collisions on them. In a very simple isothermal description ($\delta T_e = 0$), we then arrive at the density response in equation (3.16). Contrary to equation (5.133) it leads to an MHD-type response for large collisionality. This means that collisions prevent the electrons from moving along the field lines. When the electron temperature perturbations are included, the Braginskii equations lead to an electrostatic parallel electron current of the form

$$j_{\parallel e} = enD_e\hat{\mathbf{e}}_\parallel \cdot \left(\frac{1}{n}\nabla_\parallel n - \frac{e}{T_e}\nabla_\parallel\phi + 1.71\frac{\nabla_\parallel T_e}{T_e}\right) \tag{5.214}$$

where $D_e = T_e/(0.5m_e\nu_{ei})$. In the electron energy equation we now have to include the contribution from $\kappa_{\parallel e}$ in addition to $\boldsymbol{q}_{*e}$ in equation (2.26). The perpendicular collisional heat flow is always smaller than $\boldsymbol{q}_*$ as long as $\nu_e \ll \Omega_{ce}$. We then obtain

$$\frac{\delta T_e}{T_e} = \frac{1}{\omega - 5\omega_{De}/3 + i\nu_T}\left[\frac{2}{3}\omega\frac{\delta n_e}{n_e} + \left(\eta_e - \frac{2}{3}\right)\omega_{*e}\frac{e\phi}{T_e}\right] \tag{5.215}$$

where $\nu_T = 1.06k_\parallel^2 D_e$.

In the absence of ν_T, equation (5.215) is of the same form as equation (5.172) for ions, and exactly the electron temperature perturbation used for collisionless trapped electrons in the derivation of equation (5.181). In equation (5.215), however, the two-dimensional expression is recovered for very strong collisions since electrons are prevented from moving along the field lines by collisions. We may also point out that fluid closures that make use of kinetic wave–particle resonances can be obtained by a suitable choice of ν_T. In the collisionless regime, we recover the isothermal limit. By using equations (5.215), (5.214) and the electron continuity equation, we can now derive the electron density perturbation in the form

$$\frac{\delta n_e}{n_e} = \frac{\omega(\omega_{*e} - \omega_{De}) + EEH\omega_{*e}\omega_{De} + \mathrm{i}k_{\parallel}^2 D_e T(\omega)}{N_e + \mathrm{i}k_{\parallel}^2 D_e \hat{N}(\omega)} \tag{5.216}$$

where N_e is given by equation (5.182), and

$$\begin{aligned} EEH &= \eta_e - \tfrac{7}{3} + \tfrac{5}{3}\varepsilon_n \\ T(\omega) &= \omega - \tfrac{5}{3}\omega_{De} - 1.71\omega_{*e}(\eta_e - \tfrac{2}{3}) + 1.06(\omega_{*e} - \omega_{De}) \\ \hat{N}(\omega) &= \omega - \tfrac{5}{3}\omega_{De} + 1.14\omega + 1.06(\omega - \omega_{De}). \end{aligned}$$

For ions we use our previous reactive drift wave description, i.e., equation (5.173).

In order to obtain an eigenvalue equation we now make the replacement

$$k_{\parallel}^2 \rightarrow -\frac{1}{q^2R^2}\frac{\partial^2}{\partial\theta^2}.$$

Since we shall only consider the strong ballooning case here, $k_{\parallel}$ will not operate on $g(\theta)$ and $k_{\perp}(\theta)$, as given by equations (5.203) and (5.204). We can then obtain our eigenvalue equation directly from equations (5.216) and (5.173) by letting $k_{\parallel}$ operate only on ϕ, and we have no problem with non-commuting operations. This eigenvalue equation will, however, be of fourth order in general.

Since in the following we shall be considering only the strong ballooning approximation, we shall neglect the fourth order operator, thus arriving at a second order equation of the form

$$G\frac{D_e}{q^2R^2}\frac{\partial^2\phi}{\partial\theta^2} = -\mathrm{i}(A_4\omega^4 + A_3\omega^3 + A_2\omega^2 + A_1\omega + A_0)\phi \tag{5.217}$$

where

$$\begin{aligned} G = {} & (\omega - \tfrac{5}{3}\omega_{De})D - [\omega_{*e}(1.71\eta_e - 1.73) - 1.06\omega_{De}]N_i \\ & - (2.20\omega - 1.06\omega_{De})[\omega\omega_{*e}(1 - \varepsilon_n) + EIH\omega_{*e}\omega_{Di}] \end{aligned} \tag{5.218}$$

where N_i is given by equation (5.182), and

$$D = \omega^2 - \omega\left[\omega_{*e} - \omega_{De}\left(1 + \frac{10}{3\tau}\right)\right] - EIH\omega_{*e}\omega_{Di} + k_{\perp}^2\rho_s^2\omega(\omega - \omega_{*iT})$$

is the local dispersion function for the toroidal η_i mode, where

$$EIH = \eta_i - \tfrac{7}{3} + \tfrac{5}{3}\varepsilon_n.$$

Furthermore

$$
\begin{aligned}
A_4 &= k_\perp^2\rho_s^2\\
A_3 &= -k_\perp^2\rho_s^2[\omega_{*iT} + \tfrac{5}{3}\omega_{Di}(1-2\tau)]\\
A_2 &= \omega_{*e}\omega_{De}\left\{EEH + \frac{1}{\tau}EIH + \frac{10}{3}(1-\varepsilon_n)\left(1+\frac{1}{\tau}\right)\right.\\
&\quad \left. + \frac{5}{3\tau}k_\perp^2\rho_s^2\left[\frac{1+\eta_i}{\tau} - 2\left(1+\eta_i+\frac{5}{3}\varepsilon_n\right) + \varepsilon_n\tau\right]\right\}\\
A_1 &= \frac{5}{3}\omega_{*e}\omega_{De}^2\left\{\frac{2}{\tau}(EEH - EIH) - (1-\varepsilon_n)\left(1-\frac{1}{\tau^2}\right)\right.\\
&\quad \left. - k_\perp^2\rho_s^2\left[\frac{10}{3\tau^2}(1+\eta_i) - \frac{1}{\tau}(1+\eta_i) - \frac{5}{3\tau}\varepsilon_n\right]\right\}\\
A_0 &= \tfrac{5}{3}\omega_{*e}\omega_{De}\omega_{Di}^2(EEH + \tau EIH) + \tfrac{25}{9}k_\perp^2\rho_s^2\omega_{De}^2\omega_{Di}\omega_{*iT}. \qquad (5.219)
\end{aligned}
$$

We note that collisions only enter through D_e in the operator. The local limit of equation (5.217), corresponding to neglecting the operator, is thus identical to the dispersion relation (5.181) in the limit $f_t = 1$. We also note that in the edge region we actually have $\varepsilon_n \ll 1$, and in connection with strong heating we expect $\eta_i, \eta_e \gg 1$. It is thus interesting to consider the two orderings $\omega \sim \omega_*$ and $\omega \sim \omega_D$ discussed in section 5.11.5. For large ω we have the resistive ballooning mode.

5.12.1 Resistive ballooning mode

Resistive ballooning modes have been studied for a long time, both in electromagnetic and electrostatic models. These modes have generally been weakly ballooning with small growth rates. It was recently found [5.159] that such modes are stable when the shear parameter approaches 1. This effectively rules out this mode as a candidate for edge transport. In the same work, the presence of a strongly ballooning mode was also noted. This mode has a growth rate of the ideal MHD order, and is thus a very strong candidate for explaining edge transport. The work by Novakovskii *et al* [5.159], however, ignored temperature perturbations, which we expect to be very important at the edge in connection with strong heating.

We shall here, for simplicity, use the approach of [5.172] and ignore electron temperature perturbations. This greatly simplifies the algebra at the same time as retaining the important effect of ion temperature gradient on the FLR stabilization (compare equation (5.194)). Then, including the same geometry as for ion temperature gradient driven modes in equations (5.203) and (5.204), we obtain an eigenvalue equation of the same form as equation (5.206). The resulting dispersion relation is

$$
\omega(\omega - \omega_{*iT} + \mathrm{i}\gamma_D) = \frac{\omega_{*e}\omega_{Di}}{k_\theta^2\rho_s^2}(1+\tau+\eta_i) \qquad (5.220)
$$

where

$$\gamma_D = \frac{|s|}{k_\theta \rho_s}\sqrt{-\mathrm{i}(\omega - \omega_{*iT})\left(1 - \frac{\omega_{*e}}{\omega}\right)\frac{D_e}{q^2 R^2}}. \tag{5.221}$$

Here γ_D acts as a shear damping. The right-hand side of equation (5.220) gives an ideal MHD growth rate. When it is fully developed, equation (5.221) can be further simplified. In this limit the exponent α of the eigenfunction, as defined in equation (5.208), can be written

$$\alpha \approx \frac{|s|}{2} q (k_\theta \rho_s)^{1/2} \sqrt{(\varepsilon_n \Gamma)^{1/2} \frac{\omega_* \nu_{ei}}{v_{\mathrm{th}\,e}^2 / R^2}} \tag{5.222}$$

where $\Gamma = 1 + (1 + \eta_i)/2$.

We may here realistically estimate the root to be of order 1. For $s \approx 1$ we remember that $g(\theta)$ is close to 1 for $\theta < 3$. The condition for the strong ballooning regime is then $\alpha\theta^2 \geq 1$, i.e., $\alpha \geq 1/9$. This condition is easily fulfilled by equation (5.222). In deriving equation (5.220), we neglected ε_n^2 terms since ε_n is small at the edge, and we have been considering frequencies of order ω_* or larger. This means that we could have used a simpler fluid model, ignoring $\nabla \cdot \boldsymbol{q}_*$. Numerical investigations have shown that the strongly ballooning resistive ballooning mode has its maximum growth rate around

$$k_\theta \rho_s \approx 0.15.$$

Below this value the convective damping is the dominant stabilizing mechanism, and above this value the FLR stabilization becomes more important. In the local limit the condition for FLR stabilization is

$$\frac{1}{4}\omega_*^2 (1 + \eta_i)^2 \geq \frac{\omega_{*i}\omega_{Di}}{(k_\theta \rho_s)^2}\tau\Gamma. \tag{5.223}$$

With $\varepsilon_n \sim (k_\theta \rho_s)^2$ this condition leads to the stability condition

$$\eta_i \geq 3. \tag{5.224}$$

We note that if electron temperature gradients are included, Γ generalizes to $1 + \eta_e + (1 + \eta_i)/\tau$, which is the combination appearing in the MHD stability parameter α in equation (5.62). This leads to a slight increase in the threshold (5.224). It is interesting to compare the threshold (5.224) to that for stabilization due to a poloidal sheared rotation. The neoclassical poloidal rotation v_θ is of the order

$$v_{\theta i} \sim \eta_i v_{*i}. \tag{5.225}$$

A crude estimate of saturation, which usually overestimates the effect of rotation, is

$$\left|\frac{\mathrm{d}v_\theta}{\mathrm{d}r}\right| \geq \gamma \tag{5.226}$$

where γ is the growth rate in the absence of rotation. A natural estimate of $\mathrm{d}v_\theta/\mathrm{d}r$ for steep temperature gradients is

$$\frac{\mathrm{d}v_\theta}{\mathrm{d}r} \approx \frac{v_\theta}{L_T}. \tag{5.227}$$

Then equation (5.226) leads to the condition

$$\eta_i \geq k_\theta L_{Ti}\sqrt{\frac{\varepsilon_n}{(k_\theta \rho_i)^2}\tau\Gamma}. \tag{5.228}$$

Since the root is here typically of order 1, equation (5.228) expresses the fact that stabilization by rotation at reasonably moderate η requires an L_T of the order of the wavelength of the perturbation. Now, since $L_{Ti} = 0.5\varepsilon_n R/\eta_i$, we obtain for $\varepsilon_n \sim (k_\theta \rho_i)^2$

$$\eta_i \geq (0.5\varepsilon_n k_\theta R)^{\frac{2}{3}}. \tag{5.229}$$

In equation (5.229) we kept only the η_i part of Γ. For typical edge parameters, equation (5.229) gives a threshold $\eta_i \geq 7$. This threshold would also increase somewhat if we included the electron temperature gradient.

Thus, in conclusion we note that there is a strong ballooning resistive mode for edge parameters with a maximum growth rate of the ideal MHD magnitude. This mode is further stabilized by temperature gradients for $\eta_i \leq$ 2–3 and stabilized for $\eta_i \sim$ 3–5. This mode is the most likely cause of strong edge transport observed in experiments. We also note that the FLR stabilization of this mode in the local limit is also described by equation (5.181) for $f_t = 1$. We can thus extend the applicability of equation (5.181) if we reinterpret f_t as the fraction of electrons that do not move along the magnetic field, due to the combined influence of trapping and resistivity. We can, in fact, extend the picture to also include the effect of magnetic induction, which also reduces electron motion along the field lines. We may broadly say that equations (5.180) and (5.181) describe the transition from drift-type modes for $f_t = 0$ to MHD-type modes for $f_t \to 1$.

We also note that the FLR stabilization for large η_i typically appears to occur for smaller η_i than the stabilization due to a sheared poloidal rotation. In a scenario when a transition to an enhanced confinement regime is caused by increasing temperature gradients in connection with increasing heating power, we expect the FLR stabilization to set in before the stabilization due to a sheared rotation.

5.12.2 Transport in the enhanced confinement state

When the resistive ballooning mode described by equation (5.220) is stable, transport in the system described by equation (5.217) is strongly reduced, corresponding to an enhanced confinement state. In this regime the relevant ordering of ω is $\omega \sim \omega_D$. With this ordering and $\varepsilon_n \ll 1$, $\varepsilon_n\eta \sim 1$, $k_\theta^2\rho_s^2 \sim \varepsilon_n$

and $k_\perp^2 \rho_s^2 \sim \varepsilon_n^{1/2}$, we obtain an eigenvalue equation which is cubic in ω, but which becomes quadratic in the local limit. Again using the geometry defined by equations (5.203) and (5.204) with the solution in equation (5.209), we obtain the dispersion equation

$$\omega^2 - \tfrac{10}{3}\xi\omega_{De}\omega + \tfrac{5}{3}\delta\omega_{De}^2 = \mathrm{i}\gamma_D^2 \tag{5.230}$$

where

$$\xi = \frac{\eta_i - \eta_e + (\tau/2) - (1/2\tau)}{\eta_i + \tau\eta_e + 1 + \tau} \qquad \delta = \frac{\eta_i + (1/\tau)\eta_e - 7(1 + (1/\tau)/3}{\eta_i + \tau\eta_e + 1 + \tau}$$

and

$$\gamma_D^2 = k_\theta \rho_s \frac{|s|}{q}\frac{\eta_i}{\eta_i + \tau\eta_e}\sqrt{\frac{5}{3}\mathrm{i}FH\frac{D_e}{R^2\omega_{De}}} \tag{5.231}$$

$$F = \omega_{De}(3.2\omega - 2.7\omega_{De}) - 1.7\tau\frac{\eta_e}{\eta_i}\left(\omega^2 - \frac{10}{3}\omega_{Di}\omega + \tfrac{5}{3}\omega_{Di}^2\right)$$

$$H = \left(2 - \frac{1}{\tau}\right)\omega^2 + \left(\frac{10}{3\tau} - 1\right)\omega\omega_{De} - \frac{5}{3\tau}\omega_{De}^2.$$

We have here kept terms of order 1 in the local part since these are actually going to determine the threshold in many cases. In the non-local parts we have, however, strictly ignored terms of order 1 as compared to η. The relative complexity of equation (5.231) is due to the fact that we also kept the electron temperature perturbations and gradients. Here γ_D represents the effects of electron motion along the field lines and the left-hand, local part of equation (5.230) reduces to equation (5.195) for $f_t = 1$. We note that for $D_e/R^2 \approx \omega_{De}$, γ_D^2 is typically considerably smaller than ω, so that the local approximation is valid. The mode profile is determined by

$$\alpha = \frac{1}{2}k_\theta\rho_s|s|q\sqrt{\frac{5}{3}\mathrm{i}\frac{H}{F}\frac{R^2\omega_{De}}{D_e}}. \tag{5.232}$$

Again we note that for $s \approx 1$ we need $\alpha > 1/9$ for the strong ballooning approximation to be valid. This is easy to fulfil for typical edge parameters. The local part of equation (5.230) has the solution

$$\omega = \tfrac{5}{3}\xi\omega_{De} \pm \tfrac{5}{3}\omega_{De}\sqrt{\xi^2 - \tfrac{3}{5}\delta}. \tag{5.233}$$

A necessary condition for instability is clearly

$$\eta_i + \frac{1}{\tau}\eta_e > \frac{7}{3}\left(1 + \frac{1}{\tau}\right). \tag{5.234}$$

$$\frac{\delta T_e}{T_e} = \frac{\eta \omega_*}{\omega - \frac{5}{3}\omega_{De}} \frac{e\phi}{T_e}$$

compression

convection

ϕ $\qquad$ δn

density response

$$\frac{\delta n_j}{n} = \frac{\eta_j \omega_{Dj}}{\omega^2 - \frac{10}{3}\omega_{Dj}\omega + \frac{5}{3}\omega_{Dj}^2} \frac{e\phi}{T_e}$$

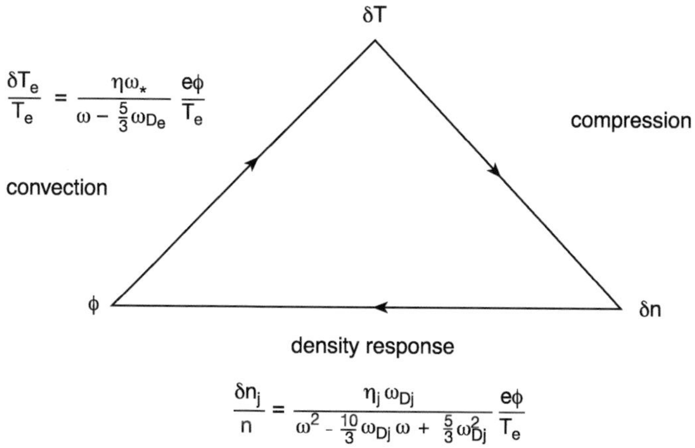

Figure 5.14. Feedback loop for the condensation instability.

This mode is of an MHD character corresponding to $k_\parallel = 0$. It can be obtained from the condition $\nabla \cdot j_\perp = 0$, which, when we neglect FLR effects, is of the form

$$\delta P = 0. \tag{5.235}$$

Here δP is the total perturbed electron plus ion pressure. It means that the density is larger where the temperature is lower and equation (5.233) is thus a kind of *condensation instability*. This mode is symmetric in ion and electron quantities and has its largest growth rate when $\eta_i = \eta_e$. It cannot be stabilized for large temperature gradients since ξ has η to the same power in numerator and denominator. The reason for this is that the main contributions to the density perturbations come from the convective temperature perturbations through compressibility. This means that a given density perturbation gives rise to a potential perturbation which is inversely proportional to η. This relation replaces the Boltzmann relation in the feedback loop of thermal instabilities (figure 5.14). Another consequence of this is that ω would cancel out after taking the final convective temperature perturbation in the feedback loop, unless we include the effect of the heat flow. This would remove the phase shift necessary for instability. *Thus the diamagnetic heat flux is required for instability.*

Finally we repeat what was pointed out in the discussion following equation (5.184), that this mode always produces a particle pinch. The transport coefficients in equations (5.186)–(5.188) are clearly valid in the local limits for both the resistive ballooning mode and the condensation mode if we take $f_t = 1$. A tendency for a particle pinch at the edge has been seen in several H-mode plasmas [5.105]. We finally note that an H-mode transition was obtained dynamically in predictive simulations using the transport coefficients (5.186)–(5.188) [5.114]. This was obtained for $f_t \approx 0.65$, in which case an η_i of 15 was needed for the transition. In our present resistive system we expect the transition to occur instead at $\eta_i \approx 5$.

5.13 Discussion

We have, in the present chapter, extended the theories of chapters 3 and 4 to more realistic geometries. This gave rise to eigenvalue equations that were solved both for some drift-type modes and for some MHD-type modes. We have also included temperature gradient driven modes. Drift kinetic and gyrokinetic equations which apply to realistic geometries have been derived, and a transport model based on an advanced fluid model for η_i and trapped electron modes has been presented. The general closure problem for fluid models has also been discussed in some detail. Finally, we have included a section on resistive edge modes, where a mechanism for the L to H mode transition has also been suggested. A condensation mode, able to give a particle pinch in the H mode, was also presented. The present chapter essentially shows the present state of research on transport, while the MHD parts are mainly included for educational and reference purposes.

5.14 Exercises

1. Use the simple trial function $\phi = 1$ in the quadratic form of equation (5.61) for $\Omega^2 = 0$. Compare the result with equation (5.62).
2. Use equation (5.133), where ν_{eff} is neglected for electrons, and include a gravity force for the ions to show that interchange modes can be due to electron trapping.
3. Generalize exercise 4 in chapter 2 by including curvature effects to a fluid description.
4. Generalize exercise 4 in chapter 2 by using the drift kinetic equation (5.105), assuming a Maxwell distribution.
5. Derive equation (5.97).
6. Show that equation (3.70) is unchanged in the presence of an electron temperature gradient when the equation of state (2.33) is used.

References

[5.1] Rudakov L I and Sagdeev R Z 1961 *Sov. Phys.–Dokl.* **6** 415
[5.2] Krall N A and Rosenbluth M N 1962 *Phys. Fluids* **5** 1435
[5.3] Kulsrud R M 1963 *Phys. Fluids* **6** 904
[5.4] Furth H P, Killeen J and Rosenbluth M N 1963 *Phys. Fluids* **4** 459
[5.5] Krall N A and Rosenbluth M N 1965 *Phys. Fluids* **8** 1488
[5.6] Kadomtsev B B 1965 *Plasma Turbulence* (New York: Academic)
[5.7] Coppi B, Laval G, Pellat R and Rosenbluth M N 1966 *Nucl. Fusion* **6** 261

[5.8] Rosenbluth M N, Zagdeev R Z, Taylor J B and Zaslavsky G M 1966 *Nucl. Fusion* **6** 297
[5.9] Liu C S 1969 *Phys. Fluids* **12** 1489
[5.10] Kadomtsev B B and Pogutse O P 1969 *Sov. Phys.–Dokl.* **14** 470
[5.11] Rutherford P H and Frieman E A 1968 *Phys. Fluids* **11** 569
[5.12] Pearlstein L D and Berk H L 1969 *Phys. Rev. Lett.* **23** 220
[5.13] Kadomtsev B B and Pogutse O P 1970 *Reviews of Plasma Physics* vol 5, ed M A Leontovitch (New York: Consultants Bureau) p 249
[5.14] Kadomtsev B B and Pogutse O P 1971 *Nucl. Fusion* **11** 67
[5.15] Rosenbluth M N and Sloan M N 1971 *Phys. Fluids* **14** 1725
[5.16] Frieman E A 1970 *Phys. Fluids* **13** 490
[5.17] Catto P J, El Nadi A M, Liu C S and Rosenbluth M N 1974 *Nucl. Fusion* **14** 405
[5.18] Yoshikawa S and Okabayashi M 1974 *Phys. Fluids* **17** 1762
[5.19] Coppi B and Rewoldt G 1974 *Phys. Rev. Lett.* **22** 1329
[5.20] D'Ippolito D A and Davidson R C 1975 *Phys. Fluids* **18** 1507
[5.21] Rozhanskii V A 1981 *JETP Lett.* **34** 56
[5.22] Rosenbluth M N and Tang W M 1976 *Phys. Fluids* **19** 1040
[5.23] Adam J C, Tang W M and Rutherford P H 1976 *Phys. Fluids* **19** 1561
[5.24] Ross D W, Tang W M and Adam J C 1977 *Phys. Fluids* **20** 613
[5.25] Hatori T, Lee Y C and Tange T 1977 *J. Phys. Soc. Japan* **43** 655
[5.26] Taylor J B 1977 *Plasma Physics and Controlled Nuclear Fusion Research* vol 2 (Vienna: IAEA) p 323
[5.27] Hasselberg G, Rogister A and El-Nadi A 1977 *Phys. Fluids* **20** 982
[5.28] Coppi B and Pegoraro F 1977 *Nucl. Fusion* **17** 969
[5.29] Coppi B 1977 *Phys. Rev. Lett.* **39** 939
[5.30] Dobrott D, Nelson D B, Greene J M, Glasser A H, Chance M S and Frieman E A 1977 *Phys. Rev. Lett.* **39** 943
[5.31] Gaudreau M *et al* 1977 *Phys. Rev. Lett.* **39** 1266
[5.32] Cordey J G and Hastie R J 1977 *Nucl. Fusion* **17** 523
[5.33] Chen L, Hsu J, Kaw P K and Rutherford P H 1978 *Nucl. Fusion* **18** 137
[5.34] Callen J D 1977 *Phys. Rev. Lett.* **39** 1540
[5.35] Antonsen T M Jr 1978 *Phys. Rev. Lett.* **41** 33
[5.36] Tang W M 1978 *Nucl. Fusion* **18** 1089
[5.37] Mikhailovskii A B 1978 *Sov. J. Plasma Phys.* **4** 683
[5.38] Wesson J A 1978 *Nucl. Fusion* **18** 87
[5.39] Connor J W, Hastie R J and Taylor J B 1978 *Phys. Rev. Lett.* **40** 396
[5.40] Lortz D and Nührenberg J 1978 *Phys. Rev. Lett.* A **68** 49
[5.41] Rechester A B and Rosenbluth M N 1978 *Phys. Rev. Lett.* **40** 38
[5.42] Connor J W, Hastie R J and Taylor J B 1979 *Proc. R. Soc.* A **365** 1
[5.43] Coppi B, Filreis J and Pegoraro F 1979 *Ann. Phys., NY* **121** 1
[5.44] Coppi B, Ferreira A, Mark J W K and Ramos J J 1979 *Nucl. Fusion* **19** 715
[5.45] Mercier C 1979 *Plasma Physics and Controlled Nuclear Fusion Research: Proc. 7th Int. Conf. (Innsbruck, 1978)* vol 1 (Vienna: IAEA) p 701
[5.46] Strauss H R 1979 *Phys. Fluids* **22** 1079
[5.47] Ross D W 1979 *Phys. Fluids* **22** 1215
[5.48] Ross D W and Mahajan S M 1979 *Phys. Fluids* **22** 294
[5.49] Gladd N T and Liu C S 1979 *Phys. Fluids* **22** 1289

[5.50] Catto P J, Rosenbluth M N and Tsang K T 1979 *Phys. Fluids* **22** 1284
[5.51] Chu C, Chu M S and Ohkawa T 1978 *Phys. Rev. Lett.* **41** 653
[5.52] Manheimer W M and Antonsen T M Jr 1979 *Phys. Fluids* **22** 957
[5.53] Antonsen T M and Lane B 1980 *Phys. Fluids* **23** 1205
[5.54] Sanuki H, Watanabe T and Watanabe M 1980 *Phys. Fluids* **23** 158
[5.55] Chen L and Cheng C Z 1980 *Phys. Fluids* **23** 356
[5.56] Tang W M, Connor J W and Hastie R J 1980 *Nucl. Fusion* **20** 1439
[5.57] Drake J F, Gladd N T, Liu C S and Chang C L 1980 *Phys. Rev. Lett.* **44** 994
[5.58] Gladd N T, Drake J F, Chang C L and Liu C S 1980 *Phys. Fluids* **23** 1182
[5.59] Rosenbluth M N 1981 *Phys. Rev. Lett.* **46** 1525
[5.60] Mondt J P 1981 *Phys. Fluids* **24** 1279
[5.61] Horton W, Choi D I and Tang W M 1981 *Phys. Fluids* **24** 1077
[5.62] Cheng C Z and Tsang K T 1981 *Nucl. Fusion* **21** 643
[5.63] Hazeltine H D, Hitchcock D A and Mahajan S M 1981 *Phys. Fluids* **24** 180
[5.64] Dewar R L, Manickam J, Grimm R C and Chance M S 1981 *Nucl. Fusion* **21** 493
[5.65] Greene J M and Chance M S 1981 *Nucl. Fusion* **21** 453
[5.66] Antonsen T M Jr, Fereira A and Ramos J J 1982 *Plasma Phys.* **24** 197
[5.67] Wagner F *et al* 1982 *Phys. Rev. Lett.* **53** 1453
[5.68] Weiland J and Mondt J P 1982 *Phys. Rev. Lett.* **48** 23
[5.69] Itoh K, Inoue-Itoh S, Tokuda S and Tuda T 1982 *Nucl. Fusion* **22** 1031
[5.70] Cheng C Z 1982 *Phys. Fluids* **25** 1020
[5.71] Andersson P and Weiland J 1983 *Phys. Rev.* A **27** 1556
[5.72] Rosenbluth M N, Tsai S T, van Dam J W and Engquist M G 1983 *Phys. Rev. Lett.* **51** 1967
[5.73] Hasselberg G and Rogister A 1983 *Nucl. Fusion* **23** 1351
[5.74] Guzdar P N, Chen L, Tang W M and Rutherford P H 1983 *Phys. Fluids* **26** 673
[5.75] Greenwald M *et al* 1984 *Phys. Rev. Lett.* **53** 352
[5.76] Chen L, White R B and Rosenbluth M N 1984 *Phys. Rev. Lett.* **52** 1122
[5.77] White R B, Chen L, Romanelli F and Hay R 1985 *Phys. Fluids* **18** 278
[5.78] Weiland J and Chen L 1985 *Phys. Fluids* **28** 1359
[5.79] Friberg J P and Wesson J A 1985 *Nucl. Fusion* **25** 759
[5.80] Grimm R C *et al* 1985 *Nucl. Fusion* **25** 805
[5.81] Andersson P and Weiland J 1985 *Nucl. Fusion* **25** 1761
[5.82] Liewer P C 1985 *Nucl. Fusion* **25** 543
[5.83] Dominguez R R and More R R 1986 *Nucl. Fusion* **26** 85
[5.84] Andersson P and Weiland J 1986 *Phys. Fluids* **29** 1744
[5.85] Rogister A, Hasselberg G, Kaleck A, Boileau A, Van Andel H W H and von Hellerman M 1986 *Nucl. Fusion* **26** 797
[5.86] Romanelli F, Tang W M and White R B 1986 *Nucl. Fusion* **26** 1515
[5.87] Dominguez R R and Waltz R E 1987 *Nucl. Fusion* **27** 65
[5.88] Thoul A A, Similon P L and Sudan R N 1987 *Phys. Rev. Lett.* **59** 1448
[5.89] Tang W M, Rewoldt G and Chen L 1986 *Phys. Fluids* **29** 3715
[5.90] Andersson P and Weiland J 1988 *Phys. Fluids* **31** 359
[5.91] Jarmén A, Andersson P and Weiland J 1987 *Nucl. Fusion* **27** 941
[5.92] Andersson P 1986 *PhD Thesis, Technical Report no 167* Chalmers University of Technology, Göteborg

[5.93] Weiland J and Nordman H 1988 *Theory of Fusion Plasmas: Proc. Varenna–Lausanne Workshop (Chexbres, 1988)* (Bologna: Editrice Compositori) p 451
[5.94] Nordman H and Weiland J 1988 *Theory of Fusion Plasmas: Proc. Varenna–Lausanne Workshop (Chexbres, 1988)* (Bologna: Editrice Compositori) p 459
[5.95] Weiland J 1988 *Comment. Plasma Phys. Control. Fusion* **12** 45
[5.96] Rogister A, Hasselberg G, Waelbroeck F and Weiland J 1988 *Nucl. Fusion* **28** 1053
[5.97] Horton W, Hong B G and Tang W M 1988 *Phys. Fluids* **31** 2971
[5.98] Dominguez R and Waltz R 1988 *Phys. Fluids* **31** 3147
[5.99] Coppi B 1988 *Phys. Lett.* A **128** 193
[5.100] Nordman H and Weiland J 1989 *Nucl. Fusion* **29** 251
[5.101] Biglari H, Diamond P H and Rosenbluth M N 1989 *Phys. Fluids* B **1** 109
[5.102] Romanelli F 1989 *Phys. Fluids* B **1** 1018
[5.103] Weiland J, Jarmén A and Nordman H 1989 *Nucl. Fusion* **29** 1810
[5.104] Hong B G, Horton W and Choi D I 1989 *Phys. Fluids* B **1** 1589
[5.105] Hamaguchi S and Horton W 1990 *Phys. Fluids* **2** 1833
[5.106] Stambaugh R D *et al* 1990 *Phys. Fluids* B **2** 2941
[5.107] Rewoldt G and Tang W M 1990 *Phys. Fluids* B **2** 318
[5.108] Nilsson J, Liljeström M and Weiland J 1990 *Phys. Fluids* B **2** 2568
Nilsson J, Liljeström M and Weiland J 1988 *Chalmers University of Technology Preprint* CTH-IEFT/PP-1998-14
[5.109] Nordman H, Weiland J and Jarmén A 1990 *Nucl. Fusion* **30** 983
[5.110] Scott S D *et al* 1990 *Phys. Rev. Lett.* **64** 531
[5.111] Romanelli F and Briguglio S 1990 *Phys. Fluids* B **2** 754
[5.112] Terry P W and Diamond P H 1990 *Phys. Fluids* B **2** 1128
[5.113] Kingsbury O T and Waltz R E 1991 *Phys. Fluids* B **3** 3539
[5.114] Weiland J and Nordman H 1991 *Nucl. Fusion* **31** 390
[5.115] Shukla P K and Weiland J 1989 *Phys. Lett.* A **137** 132
[5.116] Hammet G W and Perkins F W 1990 *Phys. Rev. Lett.* **64** 3019
[5.117] Hammet G W, Dorland W and Perkins F W 1992 *Phys. Fluids* B **4** 2052
[5.118] Cowley S C, Kulsrud R M and Sudan R 1991 *Phys. Fluids* B **3** 2767
[5.119] Garbet X, Mourges F and Samain A 1990 *Plasma Phys. Control. Fusion* **32** 917
[5.120] Kadomtsev B B 1991 *Nucl. Fusion* **31** 1301
[5.121] Weiland J and Hirose A 1992 *Nucl. Fusion* **32** 151
[5.122] Weiland J 1992 *Phys. Fluids* B **4** 1388
[5.123] Christiansen J P *et al* 1992 *Plasma Phys. Control. Fusion* **34** 1881
[5.124] Itoh K, Itoh S-I and Fukuyama A 1992 *Phys. Rev. Lett.* **69** 1050
[5.125] Wootton A J, Tsui H Y W and Prager S 1992 *Plasma Phys. Control. Fusion* **44** 2023
[5.126] Dubois M A, Sabot R, Pegouriee B, Drawn H-W and Geraud A 1992 *Nucl. Fusion* **32** 1935
[5.127] Mikhailowskii A B 1992 *Electromagnetic Instabilities in an Inhomogeneous Plasma* ed E W Laing (Bristol: Institute of Physics Publishing)
[5.128] Luce T C, Petty C C and de Haas J C M 1992 *Phys. Rev. Lett.* **68** 52
[5.129] Weiland J and Nordman H 1993 *Phys. Fluids* B **5** 1669
[5.130] Heikkilä A and Weiland J 1993 *Phys. Fluids* B **5** 2043
[5.131] Guo S C and Romanelli F 1993 *Phys. Fluids* B **5** 520
[5.132] Hirose A 1993 *Phys. Fluids* **5** 230

[5.133] Nordman H and Weiland J 1993 *Phys. Fluids* B **5** 1032
[5.134] Fröjdh M, Liljeström M and Nordman H 1992 *Nucl. Fusion* **32** 419
[5.135] Jarmén A and Fröjdh M 1993 *Phys. Fluids* B **5** 4015
[5.136] Hassam A B and Lee Y C 1984 *Phys. Fluids* **27** 438
[5.137] Mondt J P and Weiland J 1991 *Phys. Fluids* B **3** 3248
Mondt J P and Weiland J 1994 *Phys. Plasmas* **1** 1096
[5.138] Nilsson J and Weiland J 1994 *Nucl. Fusion* **34** 803
[5.139] Guo S C and Romanelli F 1994 *Phys. Plasmas* **1** 1101
[5.140] Dong J Q, Horton W and Kim J Y 1992 *Phys. Fluids* B **4** 1867
[5.141] Callen J D 1992 *Phys. Fluids* B **4** 2142
[5.142] Waltz R E, Dominguez R R and Hammett G W 1992 *Phys. Fluids* B **4** 3138
[5.143] Hua D D, Xu X Q and Fowler T K 1992 *Phys. Fluids* B **4** 3216
[5.144] Connor J W and Wilson H R 1994 *Plasma Phys. Control. Fusion* **36** 719
[5.145] Connor J W, Taylor J B and Wilson H R 1993 *Phys. Rev. Lett.* **70** 1803
[5.146] Tang W M and Rewoldt G 1993 *Phys. Fluids* **5** 2451
[5.147] Romanelli F and Zonca F 1993 *Phys. Fluids* **5** 4081
[5.148] Hirose A 1993 *Phys. Fluids* B **5** 230
[5.149] Davydova T, Jovanovic D, Vranges J and Weiland J 1994 *Phys. Plasmas* **1** 809
[5.150] Hirose A, Zhang L and Elia M 1994 *Phys. Rev. Lett.* **72** 3993
[5.151] Wagner F and Stroth U 1994 *Plasma Phys. Control. Fusion* **36** 719
[5.152] Carreras B A 1992 *Plasma Phys. Control. Fusion* **34** 1825
[5.153] Weiland J 1994 *Current Topics in the Physics of Fluids* **1** 439 (*Research Trends, Trivandrum* 1994)
[5.154] Parker S E, Dorland W, Santoro R A, Beer M A, Liu Q P, Lee W W and Hammett G W 1994 *Phys. Plasmas* **1** 1461
[5.155] Waltz R E, Kerbel G D and Milovich J 1994 *Phys. Plasmas* **1** 2229
[5.156] Nilsson J and Weiland J 1995 *Nucl. Fusion* **35** 497
[5.157] Kotschenreuther M, Dorland W, Beer M A and Hammet G W 1995 *Phys. Plasmas* **2** 2381
[5.158] Nordman H, Jarmén A, Malinov P and Persson M 1995 *Phys. Plasmas* **2** 3440
[5.159] Novakovskii S V, Guzdar P N, Drake J F, Liu C S and Waelbroeck F 1995 *Phys. Plasmas* **2** 781
[5.160] Kinsey J and Bateman G 1996 *Phys. Plasmas* **3** 3344
[5.161] Haines M G 1996 *Plasma Phys. Control. Fusion* **38** 897
[5.162] Persson M, Levandovski J L V and Nordman H 1996 *Phys. Plasmas* **3** 3720
[5.163] Waltz R E, Staebler G M, Dorland W, Hammett G W, Kotschenreuther M and Konings J A 1997 *Phys. Plasmas* **4** 2482
[5.164] Mattor N and Parker S E 1997 *Phys. Rev. Lett.* **79** 3419
[5.165] Nordman H, Strand P, Weiland J and Christiansen J P 1997 *Nucl. Fusion* **37** 413
[5.166] Guo S C and Weiland J 1997 *Nucl. Fusion* **37** 1095
[5.167] Carreras B A 1997 *IEEE Trans. Plasma Sci.* **25** 1281
[5.168] Mondt J P 1996 *Phys. Plasmas* **3** 939
[5.169] Bondeson A, Benda M, Persson M and Chu M S 1997 *Nucl. Fusion* **37** 1419
[5.170] Zagorodny A, Weiland J and Jarmén A 1997 *Comment. Plasma Phys. Control. Fusion* **17** 353
[5.171] Nordman H, Strand P, Weiland J and Christiansen J P 1997 *Nucl. Fusion* **37** 413

[5.172] Singh R, Nordman H, Anderson J and Weiland J 1998 *Phys. Plasmas* **5** 3669

[5.173] Waltz R E, Dewar R L and Garbet X 1998 *Phys. Plasmas* **5** 1784

[5.174] Bateman G, Kritz A H, Kinsey J E, Redd A J and Weiland J 1998 *Phys. Plasmas* **5** 1793

[5.175] Bateman G, Kritz A H, Kinsey J E and Redd A J 1998 *Phys. Plasmas* **5** 2355

[5.176] Strand P, Nordman H, Weiland J and Christiansen J P 1998 *Nucl. Fusion* **38** 545

[5.177] Weiland J, Nordman H and Singh R 1998 *Resistive Edge Modes: A Scenario for L–H Transition due to Heat Flux (Yokohama, 1998)* IAEA-CN-69/THP2/09

[5.178] Zagorodny A and Weiland J 1998 *Ukr. J. Phys.* **43** 1402

[5.179] Davydova T A and Weiland J 1998 *Phys. Plasmas* **5** 3089

[5.180] Christiansen J P and Cordey J G 1998 *Nucl. Fusion* **38** 1757

[5.181] Rosenbluth M N and Hinton F L 1998 *Phys. Rev. Lett.* **80** 724

[5.182] Mikkelsen D R *et al* 1999 *Proc. 16th IAEA Fusion Energy Conf. (Yokohama, 1998)* (Vienna: IAEA) IAEA-CN-69/ITER P1/08

[5.183] Rosenbluth M N 1999 *Plasma Phys. Control. Fusion* **41** A99

[5.184] Jarmén A, Malinov P and Nordman H 1998 *Plasma Phys. Control. Fusion* **40** 2041

[5.185] Singh R and Weiland J 1999 *Phys. Plasmas* **6** 1397

[5.186] Zagorodny A and Weiland J 1999 *Phys. Plasmas* **6** 2359

[5.187] Redd A J, Kritz A H, Bateman G, Rewoldt G and Tang W M 1999 *Phys. Plasmas* **6** 1162

[5.188] Kinsey J E, Waltz R E and De Boo J C 1999 *Phys. Plasmas* **6** 1865

Chapter 6

Instabilities Associated with Fast Particles in Toroidal Confinement Systems

6.1 General Considerations

As mentioned in section 5.9.3, toroidal drift wave transport gives an unfavourable scaling of the energy confinement time with heating power, roughly in agreement with the empirical scaling law (1.8). It is worth observing that this scaling is obtained with a reactive fluid model, where only magnetic drift resonances of a fluid type were included. The unfavourable scaling with heating power is due partly to the scaling of transport coefficients with temperature as $T^{-3/2}$, and partly to the threshold behaviour, i.e., $(\eta_i - \eta_{i\text{th}})^{1/2}$. These are effects of a pure (ideal) heating on the bulk plasma transport and are thus independent of the heating method. We note the close analogy with Rayleigh–Benard convection in usual fluids, where the heating itself leads to convective transport.

A different but somewhat similar picture emerges when we consider how the energy is transformed into heat for a particular heating method. This process in general requires the formation of a non-Maxwellian plasma with an energetic particle population before the external energy is transformed into heat. This is regardless of whether the heating is made by neutral beams, radiofrequency waves or alpha particles in a burning plasma. The fast particle population here is either due to injected or created particles, or due to wave–particle resonances with an injected wave. In both cases we need kinetic theory to understand the details of the relaxation. The reason for the interest in the energetic particle population is that it may lead to new instabilities, which, in turn, may cause a large transport of the energetic particles. This could lead to a situation where these particles may leave the system before depositing their energy to the bulk plasma, thus reducing the efficiency of the heating method and enhancing the unfavourable scaling of confinement time with heating power given by equation (1.8).

Although instabilities caused by fast particles have been observed experimentally [6.1–6.5], the most striking example being the ‘fishbone instability’ in

PDX [6.1], the scaling (1.8) does not seem to depend strongly on the heating method as such. It may, however, depend on which channel the energy is deposited in. (For the driftwave transport coefficients given by equations (5.186)–(5.188), the worst case, P to the power $-2/3$, is obtained for equal electron and ion heating.) This indicates that so far, instabilities caused by the energetic particles have not had a strong effect on the overall energy balance. The reason for this seems to be that the anomalous increase in the transport due to fast particles has been fairly modest. As it turns out, the fast particles also have a beneficial effect on the bulk transport [6.6–6.8], which may partly compensate for an increased transport in the fast particle channel from an overall energy balance point of view. However, since instabilities caused by energetic particles are potentially harmful [6.9–6.36], and since the situation may change in large burning plasmas, such as in a reactor, an understanding of the energetic particle physics may be essential.

6.2 The Development of Research

The first theoretical studies of energetic particle effects indicated a possibility of resonance at the Alfvén frequency [6.9] and the above-mentioned stabilizing effect on eigenmodes associated with the bulk plasma [6.6–6.8]. This is a dilution effect caused by the fact that the fast particles do not take part in the bulk instabilities. This is true for MHD-type modes as well as for drift-type modes.

Later, however, fast particles were found to introduce new modes at the precession frequency of trapped fast particles [6.12–6.14]. These modes were basically of an MHD type since one-fluid equations could be used to describe the bulk plasma. The fast particles, however, destabilized a new branch at the precession frequency of the fast particles. These types of mode were, in fact, discovered experimentally as the ‘fishbone instability’ in the PDX experiment at Princeton [6.1]. While the main fishbone mode was identified as a new branch of the internal kink mode [6.12], a precursor, with a higher mode number, seemed to be due to an analogous branch of the high-n MHD ballooning mode [6.14]. Since these modes are driven by resonant fast particles, we have sources in velocity space and thus expect to have sources in the transport equations of the type (5.166) for all fluid moments.

Experimentally, the fishbone oscillations were obtained for nearly perpendicular neutral beam injection. This led to a large trapped population of the fast particles and the instability was entirely due to the magnetic curvature drift resonance of the trapped particles. (The bounce-averaged trapped particles rotate or ‘precess’ in the toroidal direction due to magnetic curvature. The bounce-averaged magnetic drift frequency is called the precession frequency.)

The threshold in energetic particle beta of the fishbone instability is due to the continuum damping of the MHD-type mode. The MHD continuum for cylindrical plasmas was discussed at the end of section 5.4.3. For the

MHD ballooning mode the continuum damping corresponds to the iΩ term in equation (5.66). Good agreement between theory and experiment was obtained for both threshold and mode signatures for the fishbone modes.

An obvious way to reduce the effect of fishbones was to inject the neutrals more parallel to the magnetic field. This would reduce the fraction of trapped particles to below the threshold set by the continuum damping. This method, however, turned out to be only partly successful and new types of mode were excited below the continuum threshold of the fishbone instability. For these modes the Landau resonance of the circulating fast particles becomes important. As it turns out, the plasma current may cause a minimum of the Alfvén frequency as a function of radius, thus opening the possibility for modes with a discrete spectrum when the frequency is below the minimum of the Alfvén frequency. These modes were called global Alfvén eigenmodes (GAEs) [6.15, 6.16], since they could extend over the whole cross-section (compare the discussion at the end of section 5.5.3).

Another possible cause of discrete, undamped modes is toroidicity. (In fact, in a toroidal system GAE modes can also be seen as toroidal since the current enters in combination with the parallel operator $k_\parallel$ proportional to $1/R$.) The main effect of toroidicity is to couple modes with different poloidal mode numbers. This introduces a gap in the Alfvén continuum due to coupling of modes with poloidal mode numbers m and $m+1$. This is the toroidicity-induced Alfvén eigenmode, or TAE mode [6.17, 6.18]. For a while it was believed that GAE modes and TAE modes had a very small instability threshold, set only by electron Landau damping. Later, however, it was found that these modes can couple to the continuum modes by other toroidal coupling possibilities. This usually gives the main stabilizing effect.

There are other similar types of mode. We may mention EAE modes caused by the coupling between m and $m+2$ modes due to ellipticity, NAE modes which couple m and $m+3$ modes due to triangularity and BAE modes which are due to finite beta modifications of the magnetic curvature.

6.3 Dilution Due to Fast Particles

Before going into the new types of instability caused by fast particles, we shall make some general considerations of multi-ion systems, where, in particular, the effect of dilution becomes clear. We consider a system of electrons, e, main ions, i and fast ions, f. The main ions have charge 1 and the fast ions charge Z. Quasi-neutrality requires

$$n_e = n_i + Zn_f. \tag{6.1}$$

Let us now introduce the fast fraction ε_f such that

$$n_f = \varepsilon_f n_e \tag{6.2}$$

$$n_i = (1 - Z\varepsilon_f)n_e. \tag{6.3}$$

Since we require equation (6.1) to hold also for perturbations, i.e.

$$\delta n_e = \delta n_i + Z\delta n_f \tag{6.4}$$

we obtain after dividing by n_e and using equations (6.2) and (6.3)

$$\frac{\delta n_e}{n_e} = (1 - \varepsilon_f Z)\frac{\delta n_i}{n_i} + \varepsilon_f Z\frac{\delta n_f}{n_f}. \tag{6.5}$$

Equation (6.4) is the basic charge balance equation which needs to be fulfilled, regardless of which physics description we use for electrons, ions and hot ions. If we now study the influence of fast ions on a mode associated with the bulk plasma, we have $\omega_{Df} \gg \omega$. Then, using equation (5.123) or equation (5.160) for the fast ions we obtain

$$\frac{\delta n_e}{n_e} = Z\frac{e\phi}{T_e}. \tag{6.6}$$

The Boltzmann response for fast ions means that they do not take part in the destabilizing process, but just respond in an isothermal way to the potential perturbations. The instability growth rate is reduced due to the factor $1 - \varepsilon_f Z$ in front of the main ion response. This is the dilution effect. The same principle, of course, applies if we use a fast particle response which includes the parallel resonance. If we now consider modes with $\omega \sim \omega_{Df}$, the fast particles can be destabilizing. We will consider such cases in the following.

6.4 Fishbone-Type Modes

Fishbone-type modes are basically new branches of MHD-type modes, introduced by fast trapped particles. The pure MHD modes are fairly close to marginal stability, in the sense that the destabilizing effects balance the Alfvén line bending effect to give a mode with eigenfrequency close to zero. For the fishbone-type modes we are only interested in the trapped population of fast particles. For these we start from the gyrokinetic equation (5.127), and average it over the bounce motion, as described in section 5.7 for electrons. This means that the $v_\parallel$ parts vanish from equation (5.127) and the magnetic drift is replaced by the bounce-averaged precession frequency. The fast particle response is then given by equation (5.129), where the magnetic drift in the denominator is replaced by its bounce average. Now treating the fast particles as an additional particle species in the quasi-neutrality condition, we may derive a dispersion relation by the trial function method described in section 5.5.2. Ignoring the Ω^2 part, the dispersion relation (5.72) is generalized to the form

$$\mathrm{i}\Omega = \delta W_{\mathrm{MHD}} + \delta W_f. \tag{6.7}$$

Here δW_{MHD} is given by equation (5.73) for the MHD ballooning mode, and by the left-hand side of equation (5.95) for the internal kink mode. It was evaluated for the first time with toroidal effects in [6.37]. The kinetic part, δW_f, is due to the fast particle response obtained from equation (5.127). For a slowing down distribution, δW_f takes the form

$$\delta W_f = \sqrt{2}\beta_f \frac{\alpha}{\varepsilon_{nf}} \left(1 - \frac{3}{2}\varepsilon_{nh}\right) \ln\left(\frac{\alpha - 1}{\alpha}\right) \tag{6.8}$$

where β_f is the beta of the fast particles, $\alpha = \omega/\omega_{Df}$, ω_{Df} is the precession frequency and $\varepsilon_{nf} = \omega_{Df}/\omega_{*f}$, where ω_{*f} is the diamagnetic drift frequency of the fast particles. For a Maxwellian distribution we have [6.13]:

$$\delta W_f = \frac{1}{2}\beta_f \frac{\alpha}{\varepsilon_{nf}} \left\{ 2\eta_f \left[1 - W\left(\sqrt{2\alpha}\right)\right] - \alpha\varepsilon_{nf} + [\alpha\varepsilon_{nf} - 1 - \eta_f(2\alpha - 1)]\left[1 - W\left(\sqrt{2\alpha}\right)\right]^2 \right\}. \tag{6.9}$$

The difference between the distributions (6.8) and (6.9) is usually not very large. Both the real part of the eigenfrequency and the growth rate are of the order of the precession frequency. This is also true when we use the advanced reactive fluid model of section 5.9 for the fast particles [6.27]. The threshold for the reactive fluid model does, however, differ significantly. For the distributions (6.8) and (6.9) we obtain the threshold by balancing the imaginary part of δW_f with $\mathrm{i}\Omega$. For the slowing down distribution the threshold is

$$\beta_{\text{crit}} = \frac{\langle\omega_{Df}\rangle}{\omega_A} \frac{2s}{\pi^2 q^2 I_0} \tag{6.10}$$

where I_0 proportional to R/r includes the bounce averaging of the driving pressure term and $\langle\rangle$ denotes bounce averaging of the magnetic drift frequency. The growth rate is

$$\gamma = \omega_{Df} \frac{\pi^2}{4} \frac{\beta_f - \beta_{\text{crit}}}{\beta_{\text{crit}}}. \tag{6.11}$$

When the plasma is heated β_f increases until it exceeds β_{crit}. Then the fishbone mode goes unstable and rapidly reduces β_f to below the threshold. The heating again increases β_f and the process repeats itself. This leads to a fishbone-like oscillation.

6.5 Toroidal Alfvén Eigenmodes

We shall now consider the potentially most dangerous type of eigenmode, which is the discrete one. This type of mode is, in the simplest description, not subject to continuum damping. As mentioned above, the toroidal Alfvén eigenmodes (TAE modes) occur in a gap in the Alfvén continuum, caused by a coupling of

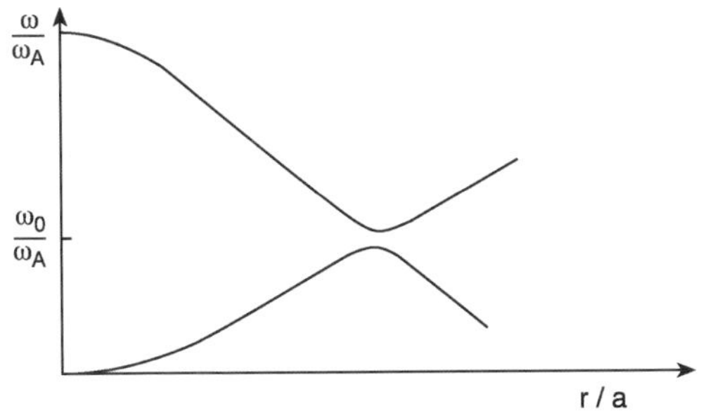

Figure 6.1. Gap due to toroidal coupling of Alfvén modes with poloidal mode numbers m and $m+1$.

modes with poloidal mode number m and $m+1$. In the cylindrical approximation $k_\parallel$ is given by equation (5.7), i.e.

$$k_\parallel = (m - nq)/qR \tag{6.12}$$

$$k_{\parallel m} = -k_{\parallel m+1}. \tag{6.13}$$

which happens when

$$q(r) = (2m+1)/2n. \tag{6.14}$$

In particular, for $m = -2$ and $n = -1$ we obtain a resonance at $q = 1.5$. For this case we show the radial continuous spectra in figure 6.1. In the gap between the full lines there is a solution with only one (discrete) ω, which is the TAE mode. It has a real eigenfrequency close to the common cylindrical eigenfrequency ω_0 of the coupled Alfvén modes. Since this frequency is large, a kinetic resonance requires fast particles. The formulation usually makes use of the condition $\nabla \cdot \boldsymbol{j} = 0$. A convenient and rather general formulation is given in [6.23] as

$$\boldsymbol{B} \cdot \nabla \frac{1}{B^2} \boldsymbol{B} \cdot \nabla\, \nabla^2 \phi + \frac{\omega(\omega - \omega_{*iT})}{v_A^2} \nabla^2 \phi - \frac{8\pi}{B^2} \frac{\mathrm{d}P}{\mathrm{d}r} k_\theta (\boldsymbol{\kappa} \times \boldsymbol{e}_\parallel) \cdot \nabla \phi$$
$$+ \frac{4\pi}{B^2} \frac{\omega}{c} \boldsymbol{\kappa} \times \boldsymbol{B} \cdot \nabla (\delta P_{\parallel f} + \delta P_{\perp f}) = 0. \tag{6.15}$$

Here $\delta P_{\parallel f}$ and $\delta P_{\perp f}$ are the parallel and perpendicular components of the hot particle pressure tensor, and the fact that only the curvature part of the magnetic drift appears is consistent with equation (5.21). In equation (6.15), the first term is the shear Alfvén line bending term, the second term comes from the divergence of the polarization and stress tensor drifts, the third term is the interchange term due to the driving bulk pressure and the fourth term is the driving term due to the pressure of the fast particles. An analytical expression for the growth rate of TAE modes, driven by only the parallel alpha particle pressure, was obtained in [6.24]:

$$\gamma = \omega_0 \frac{9}{4}\left[\beta_\alpha\left(\frac{\omega_{*\alpha}}{\omega_0} - 0.5\right)F - \beta_e \frac{v_A}{v_e}\right]. \tag{6.16}$$

Here β_α and β_e are the β values of alpha particles and electrons, respectively, and $\omega_{*\alpha}$ is the alpha particle diamagnetic frequency. $F(x) = x(1 + 2x^2 + 2x^4)\,\mathrm{e}^{-x^2}$, where $x = v_A/v_\alpha$ represents the kinetic distribution. As pointed out above, we need particles with velocity close to the Alfvén velocity for a significant growth rate.

6.6 Discussion

We have studied some of the most important collective effects associated with fast particles. These can be divided into, on the one hand, continuum modes and global modes, and on the other hand, modes driven by the perpendicular fast particle pressure through the resonance with the precession frequency of the trapped particles, or by the parallel pressure through the Landau resonance with the circulating fast ions. As it turns out, the TAE mode, which is normally excited by the transit resonance, can also be destabilized by the precessing trapped ions [6.26].

Equation (6.15) can be used to describe both fishbone type ballooning modes and TAE modes (the only missing part is the kink term). The presence of the ion diamagnetic drift in the second term means that we can also describe the kinetic ballooning mode [6.23]. For perpendicular neutral beam heating, $\delta P_{\perp f}$ will dominate and a majority of the fast ions will be trapped. For parallel neutral beam injection, $\delta P_{\parallel f}$ will dominate and the majority of the fast particles will be circulating. In the latter case we only have the TAE mode in the system. In the first case we can have both fishbone-type modes and TAE modes. A major difference between these is that the fishbone mode is triggered close to marginal stability, i.e., only close to the MHD beta limit for the ballooning type. On the other hand, it can be excited for the general fast ion precession frequency, while the TAE mode requires particles with velocity close to the Alfvén velocity.

An investigation of thresholds for fishbone-type and global modes in kinetic and reactive systems, i.e., alternatively using equation (5.129) and (5.173) for the fast particles, was made in [6.27]. Recently, more detailed kinetic calculations of the stability of TAEs for reactors [6.28, 6.33, 6.35, 6.36] indicate that the modes may be somewhat more unstable than in TFTR, although the situation should be possible to control.

Finally, the nonlinear saturation of the instabilities is, of course, fundamental for the transport they cause. A usual estimate of the saturation level is obtained by balancing the linear growth rate and the $E \times B$ trapping frequency or nonlinear frequency shift. This leads to an estimate similar to equation (3.65). Berk and Breizman have investigated the details of the kinetic saturation process, including relaxation oscillations and flattening of the distribution function [6.29–6.31, 6.33, 6.34].

References

[6.1] McGuire K M *et al* 1983 *Phys. Rev. Lett.* **50** 891

[6.2] Wong K L *et al* 1991 *Phys. Rev. Lett.* **66** 1874

[6.3] Wong K L *et al* 1992 *Phys. Fluids* B **4** 2122

[6.4] McGuire K M and the TFTR team 1995 *Phys. Plasmas* **2** 2176

[6.5] Darrow D S *et al* 1996 *Phys. Plasmas* **3** 1875

[6.6] Berk H L 1976 *Phys. Fluids* **19** 1255

[6.7] Coppi B and Pegoraro F 1979 *Comment. Plasma Phys. Control. Nucl. Fusion* **5** 131

[6.8] Weiland J 1981 *Phys. Scr.* **23** 801

[6.9] Rosenbluth M N and Rutherford P H 1975 *Phys. Rev. Lett.* **34** 1428

[6.10] Tsang K T, Sigmar D J and Whitson J C 1981 *Phys. Fluids* **24** 1508

[6.11] Rosenbluth M N, Tsai S T, van Dam J W and Engquist M G 1983 *Phys. Rev. Lett.* **51** 1967

[6.12] Chen L, White R B and Rosenbluth M N 1983 *Phys. Rev. Lett.* **51** 1967

[6.13] White R B, Chen L, Romanelli F and Hay R 1995 *Phys. Fluids* **28** 278

[6.14] Weiland J and Chen L 1986 *Phys. Fluids* **28** 1359

[6.15] Appert K, Gruber R, Troyon F and Vaclavic J 1982 *J. Plasma Phys.* **24** 1147

[6.16] Ross D W, Chen G L and Mahajan S M 1982 *Phys. Fluids* **25** 652

[6.17] Cheng C Z, Chen L and Chance M S 1985 *Ann. Phys.* **161** 21

[6.18] Cheng C Z and Chance M S 1986 *Phys. Fluids* **29** 3695

[6.19] Biglari H and Chen L 1986 *Phys. Fluids* **29** 1760

[6.20] Coppi B and Porcelli F 1986 *Phys. Rev. Lett.* **57** 2277

[6.21] Li Y M, Mahajan S M and Ross D W 1987 *Phys. Fluids* **30** 1466

[6.22] Weiland J, Lisak M and Wilhelmsson H 1987 *Phys. Scr.* T **16** 53

[6.23] Biglari H and Chen L 1991 *Phys. Rev. Lett.* **67** 3681

[6.24] Fu G Y and Van Dam J W 1989 *Phys. Fluids* B **1** 1949

[6.25] Fu G Y and Cheng C Z 1990 *Phys. Fluids* B **2** 1427

[6.26] Biglari H, Zonca F and Chen L 1992 *Phys. Fluids* B **4** 2385

[6.27] Liljeström M and Weiland J 1992 *Phys. Fluids* B **4** 630

[6.28] Candy J and Rosenbluth M N 1995 *Nucl. Fusion* **35** 1069

[6.29] Berk H L and Breizman B N 1990 *Phys. Fluids* B **2** 2235

[6.30] Berk H L and Breizman B N 1990 *Phys. Fluids* B **2** 2246

[6.31] Berk H L, Breizman B N, Fitzpatrick J, Pekker M S, Vong H V and Wong K L 1996 *Phys. Plasmas* **3** 1827

[6.32] Fülöp T, Lisak M, Kolesnichenko Ya and Anderson D 1996 *Plasma Phys. Control. Fusion* **38** 811

[6.33] Breizman B N, Candy J, Porcelli F and Berk H L 1998 *Phys. Plasmas* **5** 2326

[6.34] Breizman B N *et al* 1999 *17th IAEA Fusion Energy Conf. (Yokohama, 1998)* (Vienna: IAEA) IAEA-F1-CN-69/TH2/4

[6.35] Jaun A, Fasoli A and Heidbrink W W 1998 *Phys. Plasmas* **5** 2952

[6.36] Jaun A *et al* 1998 *Phys. Plasmas* **5** 3801

[6.37] Bussac M N, Pellat R, Edery D and Soule J L 1975 *Phys. Rev. Lett.* **35** 1638

Chapter 7

Nonlinear Theory

7.1 The Ion Vortex Equation

We have up to now studied linear and quasi-linear phenomena. Although these, in combination with an estimate of the saturation level, can be used to derive transport coefficients, it is important to go beyond this description in order to understand its region of applicability [7.1–7.81]. In particular, nonlinear cascade rules [7.18, 7.20, 7.25, 7.26, 7.29, 7.30, 7.55] are important for the interplay between sources and sinks in k-space, and the resulting saturation level and correlation length. We shall thus consider here some simple nonlinear systems for turbulence in magnetized plasmas. We shall also make a kinetic derivation of the diffusion coefficient, which involves the turbulent transport itself as a decorrelation mechanism [7.3–7.5, 7.7, 7.8]. As we have pointed out in chapter 3, the parallel ion motion may often be ignored in drift and flute modes. This is possible if $\omega \gg k_\parallel c_s$. For this case it is possible to derive a simple, but still rather general, nonlinear equation for the ion vorticity $\Omega = \nabla \times \boldsymbol{v}_i$.

We start from the fluid equation of motion for ions

$$\frac{\partial \boldsymbol{v}_i}{\partial t} + (\boldsymbol{v}_i \cdot \nabla)\boldsymbol{v}_i = \frac{e}{m_i}(\boldsymbol{E} + \boldsymbol{v}_i \times \boldsymbol{B}) - \frac{1}{n m_i}\nabla p + \boldsymbol{g}. \tag{7.1}$$

Taking the curl of this equation, and introducing

$$\boldsymbol{\Omega}_{ci} = \frac{e}{m_i}\boldsymbol{B}$$

using the Maxwell equation

$$\nabla \times \boldsymbol{E} = -\frac{\partial \boldsymbol{B}}{\partial t}$$

and the vector relation

$$(\boldsymbol{v} \cdot \nabla)\boldsymbol{v} = \frac{1}{2}\nabla v^2 - \boldsymbol{v} \times \nabla \times \boldsymbol{v}$$

we obtain

$$\frac{\partial \mathbf{\Omega}_i}{\partial t} - \nabla \times (\boldsymbol{v}_i \times \mathbf{\Omega}_i) = -\frac{\partial \Omega_{ci}}{\partial t} + \nabla \times (\boldsymbol{v}_i \times \mathbf{\Omega}_{ci}) + \frac{1}{m_i n^2} \nabla n \times \nabla p. \quad (7.2)$$

When $\nabla n \times \nabla p = 0$ we may write equation (7.2) as

$$\frac{\partial}{\partial t}(\mathbf{\Omega}_i + \mathbf{\Omega}_{ci}) = \nabla \times [\boldsymbol{v}_i \times (\mathbf{\Omega}_i + \mathbf{\Omega}_{ci})]. \quad (7.3)$$

If equation (7.3) is integrated around a closed line and we make use of Stokes' theorem, we now obtain a generalized form of the familiar theorem of attachment of magnetic field lines to the plasma, which reduces to the usual form when $\Omega_i \ll \Omega_{ci}$. In its usual form this theorem is primarily concerned with the perpendicular component of equation (7.3), while we shall here be interested only in the parallel component of equation (7.3). Now since $\nabla \cdot \mathbf{\Omega}_i = \nabla \mathbf{\Omega}_{ci} = 0$, we find

$$\nabla \times (\boldsymbol{v} \times \mathbf{\Omega}) = -\mathbf{\Omega} \nabla \cdot \boldsymbol{v} - (\boldsymbol{v} \cdot \nabla)\mathbf{\Omega}$$

where Ω represents Ω_i or Ω_{ci}. We now realize that the operator $\mathbf{\Omega} \cdot \nabla$ represents a variation in the direction of the vorticity or the magnetic field. Since the vorticity will be due mainly to the $E \times B$ drift caused by the background magnetic field, and we assume the magnetic perturbation to be small, we find that this operator represents a variation along the background magnetic field. Since the ion motion along the magnetic field is going to be neglected, we then drop this term. (The neglection of ion motion along B amounts to dropping $k_\parallel c_s$, i.e., $k_\parallel = 0$.) Then collecting terms and introducing

$$\frac{\mathrm{d}}{\mathrm{d}t} = \frac{\partial}{\partial t} + \boldsymbol{v} \cdot \nabla$$

we have

$$\frac{\mathrm{d}}{\mathrm{d}t}(\mathbf{\Omega}_i + \mathbf{\Omega}_{ci}) + (\mathbf{\Omega}_i + \mathbf{\Omega}_{ci})\nabla \cdot \boldsymbol{v}_i = \frac{1}{m_i n^2} \nabla n \times \nabla p. \quad (7.4)$$

In order to express $\nabla \cdot \boldsymbol{v}_i$ we now use the ion continuity equation, which may be written

$$\frac{\mathrm{d}n_i}{\mathrm{d}t} + n_i \nabla \cdot \boldsymbol{v}_i = 0$$

or $\nabla \cdot \boldsymbol{v}_i = -(\mathrm{d}/\mathrm{d}t) \ln n_i$. Then

$$\frac{\mathrm{d}}{\mathrm{d}t}(\mathbf{\Omega}_i + \mathbf{\Omega}_{ci}) - (\mathbf{\Omega}_i + \mathbf{\Omega}_{ci})\frac{\mathrm{d}}{\mathrm{d}t} \ln n_i = \frac{1}{m_i n^2} \nabla n \times \nabla p. \quad (7.5)$$

Since $\nabla \cdot \boldsymbol{v}_E = 0$, it is often a good approximation to drop $(\mathrm{d}/\mathrm{d}t) \ln n_i$ completely. This corresponds to incompressible flow. The equation then describes the generation of vorticity $\mathbf{\Omega}_i$ and the magnetic field $\mathbf{\Omega}_{ci}$ by the vector

$\nabla n \times \nabla p$. This vector is called the baroclinic vector and is present whenever there is an angle between the temperature and density gradient. It is one of the mechanisms responsible for the generation of magnetic fields in laser pellet experiments. For the study of the ion vortex motion it is convenient to rewrite equation (7.5) in scalar form. We then note that the nonlinear $E \times B$ drift is given by equation (1.5), i.e., for $v_{\parallel} = 0$ the nonlinear contribution disappears and $\nabla \times \boldsymbol{v} = \Omega \hat{\boldsymbol{z}}$. We thus take the scalar product of equation (7.5) with $\hat{\boldsymbol{z}}$. This means that we disregard perpendicular perturbations of $\boldsymbol{\Omega}_c$ in the ion equation. The approximation $k_{\parallel} = 0$ was also made in obtaining equation (4.31) from equation (4.20). We may rewrite equation (7.5) as

$$\frac{\mathrm{d}}{\mathrm{d}t} \ln \left(\frac{\boldsymbol{\Omega}_i + \boldsymbol{\Omega}_{ci}}{n_i} \right) = \frac{1}{m_i n^2} \frac{(\nabla n_i \times \nabla p_i) \cdot \hat{\boldsymbol{z}}}{\Omega_i + \Omega_{ci}}. \tag{7.6}$$

In order to treat consistently the ion temperature effects we have to include the velocity $\boldsymbol{v}_\pi$ due to the stress tensor. A correct evaluation of equation (7.6) in the presence of an ion temperature gradient is then rather complicated. We shall thus, for simplicity, assume the ion temperature to be small and drop the baroclinic vector. We then have the usual form of the ion vortex equation

$$\frac{\mathrm{d}}{\mathrm{d}t} \ln \left(\frac{\boldsymbol{\Omega}_i + \boldsymbol{\Omega}_{ci}}{n_i} \right) = 0. \tag{7.7}$$

We now write $n_i = n_0 + \delta n_i$, where $\delta n_i \ll n_0$, and introduce the weak nonlinearity assumption $\Omega_i \ll \Omega_{ci}$. Then equation (7.7) takes the form

$$\frac{\mathrm{d}}{\mathrm{d}t} \left(\ln \Omega_{ci} + \frac{\Omega_i}{\Omega_{ci}} - \ln n_0 - \frac{\delta n_i}{n_0} \right) = 0.$$

Since we are now going to consider the ion temperature to be small, we use $\boldsymbol{v}_i = \boldsymbol{v}_E + \boldsymbol{v}_g$ in equation (7.7). This is correct to first order in ω/Ω_{ci} since the ion inertia (polarization drift) is already included in equation (7.7). We then find

$$\Omega_i = (\nabla \times \boldsymbol{v}_E) \cdot \hat{\boldsymbol{z}} = \frac{1}{B_0} \Delta \phi. \tag{7.8}$$

We can write equation (7.7) in the form

$$\begin{aligned} &\left(\frac{\partial}{\partial t} + v_g \frac{\partial}{\partial y} \right) \left(\frac{1}{B_0 \Omega_{ci}} \Delta \phi - \frac{\delta n_i}{n_0} \right) - \frac{1}{B_0} (\hat{\boldsymbol{z}} \times \nabla \phi) \cdot \hat{\boldsymbol{x}} \frac{\mathrm{d}}{\mathrm{d}x} \ln n_0 \\ &\quad = -\frac{1}{B_0} (\hat{\boldsymbol{z}} \times \nabla \phi) \cdot \nabla \left(\frac{1}{B_0 \Omega_{ci}} \Delta \phi - \frac{\delta n_i}{n_0} \right) \end{aligned} \tag{7.9}$$

where we have dropped both time and space derivatives of Ω_{ci}. The ∇B drift due to a variation of B_0 along x may, however, be included in v_g. We have now obtained a nonlinear equation for the ion dynamics. The density perturbation

δn_i can be expressed in terms of ϕ by involving the electron dynamics and assumption of quasi-neutrality. In equation (7.9) we have dropped parallel ion motion, i.e., assumed $k_\parallel c_s \ll \omega$, which means that equation (7.9) is equivalent to the assumption $k_\parallel = 0$ for the ions. For the electrons, however, we are still free to choose the region of interest. Remaining in the drift wave interval (3.5), we can use the Boltzmann distribution (3.2) for the electrons. In combination with the quasi-neutrality condition this gives

$$\ln n_i = \ln n_0 + \frac{e\phi}{T_e}$$

which may be substituted directly into equation (7.7) without expansion. This means that for a Boltzmann distribution of electrons, the electrons will not contribute to nonlinearity. Then using the expanded form of $\ln(\boldsymbol{\Omega}_i + \boldsymbol{\Omega}_{ci})$, we obtain

$$\begin{aligned}\left(\frac{\mathrm{d}}{\mathrm{d}t} + v_g \frac{\partial}{\partial y}\right)\left(\frac{1}{B_0 \Omega_{ci}}\Delta\phi - \frac{e\phi}{T_e}\right) - \frac{\kappa}{B_0}\frac{\partial \phi}{\partial y} \\ = \frac{1}{B_0}(\nabla\phi \times \hat{z}) \cdot \nabla\left(\frac{1}{B_0\Omega_{ci}}\Delta\phi - \frac{e\phi}{T_e}\right).\end{aligned} \tag{7.10}$$

We notice that the gravitational drift in equation (7.9) only gives a Doppler shifted frequency. Thus, moving to the frame with velocity v_g we obtain the equation

$$\frac{\mathrm{d}}{\mathrm{d}t}\left(\frac{1}{B_0\Omega_{ci}}\Delta\phi - \frac{e\phi}{T_e}\right) - \frac{\kappa}{B_0}\frac{\partial\phi}{\partial y} = \frac{1}{B_0^2\Omega_{ci}}(\nabla\phi \times \hat{z}) \cdot \nabla\Delta\phi + \frac{1}{B_0 T}\frac{\mathrm{d}T_e}{\mathrm{d}x}\frac{e\phi}{T_e}\frac{\partial\phi}{\partial y}$$

where we included the possibility of an electron temperature gradient in the x direction. We may compare the two nonlinear terms in the following way. Introducing $T/eB_0 = c_s^2/\Omega_{ci}$, we can rewrite the first term as $(1/B_0)\rho^2[\nabla(e\phi/T_e) \times \hat{z}] \cdot \nabla\Delta\phi$. We then arrive at the ratio $(\mathrm{d}/\mathrm{d}x)\ln T_e/k_y^2 k^2\rho^2$ between the nonlinear terms. It is natural to assume $(\mathrm{d}/\mathrm{d}x)\ln T \ll k_y$. Since, however, $k^2\rho^2$ may be small, the second nonlinear term will not always be negligible. We shall, nevertheless assume this to be the case in the following. We then arrive at the Hasegawa–Mima equation (the quasi-geostrophic vortex equation)

$$\frac{\mathrm{d}}{\mathrm{d}t}(\rho^2\Delta\phi - \phi) - v_{*e}\frac{\partial\phi}{\partial y} = \rho^2\frac{1}{B_0}(\nabla\phi \times \hat{z}) \cdot \nabla\Delta\phi. \tag{7.11}$$

where we multiplied by T_e/e. We again observe that the interchange frequency is absent when the electrons are Boltzmann distributed. Equation (7.11) has the conserved quantities

$$W = \int [(\rho\nabla\phi)^2 + \phi^2]\mathrm{d}^3 r \tag{7.12}$$

and

$$V = \int [(\rho\nabla\phi)^2 + (\rho^2\Delta\phi)^2]\mathrm{d}^3 r \tag{7.13}$$

where W is the energy and V is the enstrophy (squared vorticity).

We notice that in the linear approximation, equation (7.11) reduces to

$$\omega = \frac{\omega_{*e}}{1 + k_y^2 \rho^2}$$

which, in the limit $k^2\rho^2 \ll 1$, reduces to equation (3.19).

We now turn to the case $\omega \gg k_{\|} v_{\mathrm{th}\,e}$. In this limit the electrons are not Boltzmann distributed. Instead, we may also use the approximation $k_{\|} = 0$ for the electrons. The electrons can then be described by the continuity equation

$$\frac{\partial n_e}{\partial t} + \frac{\kappa}{B_0}\frac{\partial \phi}{\partial y} = \frac{1}{B_0}(\nabla\phi \times \hat{z}) \cdot \nabla n_e \tag{7.14}$$

where we used $v_e = v_E$, since temperature effects do not enter the continuity equation. In the limit $\kappa \ll k$ we can write $\nabla n_e/n_0 = \nabla(\delta n_e/n_0)$. Thus, dividing equation (7.14) by n_0 and subtracting it from equation (7.9), we arrive at the coupled system of equations

$$\frac{1}{B_0\Omega_{ci}}\left(\frac{\mathrm{d}}{\mathrm{d}t} + v_g\frac{\partial}{\partial y}\right)\Delta\phi - v_g\frac{\partial}{\partial y}\frac{\delta n}{n_0} = \frac{1}{B_0^2\Omega_{ci}}(\nabla\phi \times \hat{z}) \cdot \nabla\Delta\phi \tag{7.15}$$

$$\frac{\mathrm{d}}{\mathrm{d}t}\left(\frac{\delta n}{n_0}\right) + \frac{\kappa}{B_0}\frac{\partial\phi}{\partial y} = (\nabla\phi \times \hat{z}) \cdot \nabla\frac{\delta n}{n_0} \tag{7.16}$$

where we again used the quasi-neutrality condition and introduced $\delta n \equiv \delta n_e = \delta n_i$. We notice that equation (7.15) couples to equation (7.16) due to the gravitational drift v_g alone. In the linear approximation we obtain from equation (7.16)

$$\frac{\delta n}{n_0} = \frac{\kappa}{B_0\omega}k_y\phi$$

which is the same relation as equation (3.22). Substituting this expression into equation (7.15) we find the dispersion relation

$$\omega(\omega - k_y v_g) + \kappa g k_y^2/k^2 = 0 \tag{7.17}$$

where we introduced $v_g = -g/\Omega_{ci}$ and $k^2 = k_x^2 + k_y^2$. This dispersion relation is identical to equation (3.23) if $g \to g_i + (m_e/m_i)g_e$. It is also of interest to note that when $g = v_t^2/R_c$, i.e., when it is due to the curvature, the gravitational drift of the electrons is of the same order as that of the ions for equal temperatures. For this case we should replace g by $g_i + (m_e/m_i)g_e$, which will, however, only modify the interchange frequency $(-\kappa g)^{1/2}$ by a factor $\sqrt{2}$.

The simplest nonlinear process described by equation (7.11) is the three wave interaction. We write

$$\phi = \sum \phi_k(t)\,\mathrm{e}^{\mathrm{i}(k_x x + k_y y - \omega t)}. \tag{7.18}$$

Substituting equation (7.18) into equation (7.11) we obtain

$$\left(-\mathrm{i}\omega + \frac{\mathrm{d}}{\mathrm{d}t}\right)(1 + k^2\rho^2)\phi_k - \mathrm{i}k_y v_{*e}\phi_k$$
$$= \rho^2 \frac{1}{B_0}[(\boldsymbol{k}_1 \times \hat{\boldsymbol{z}}) \cdot \boldsymbol{k}_2 k_2^2 + (\boldsymbol{k} \times \hat{\boldsymbol{z}}) \cdot \boldsymbol{k}_1 k_1^2]\phi_{k_1}\phi_{k_2}\,\mathrm{e}^{\mathrm{i}(\omega-\omega_1-\omega_2)t} \qquad (7.19)$$

where we assumed the matching condition

$$\boldsymbol{k} = \boldsymbol{k}_1 + \boldsymbol{k}_2$$

to be fulfilled. Assuming now that ω is a solution of the linear dispersion relation, we obtain the three coupled equations

$$\frac{\partial \phi_k}{\partial t} = \frac{1}{B_0}\frac{\rho^2}{1 + k^2\rho^2}(\boldsymbol{k}_1 \times \boldsymbol{k}_2) \cdot \hat{\boldsymbol{z}}(k_2^2 - k_1^2)\phi_{k_1}\phi_{k_2}\,\mathrm{e}^{\mathrm{i}\Delta\omega t} \qquad (7.20)$$

$$\frac{\partial \phi_{k_1}}{\partial t} = -\frac{1}{B_0}\frac{\rho^2}{1 + k^2\rho^2}(\boldsymbol{k} \times \boldsymbol{k}_2) \cdot \hat{\boldsymbol{z}}(k_2^2 - k^2)\phi_k\phi_{k_2}^*\,\mathrm{e}^{-\mathrm{i}\Delta\omega t} \qquad (7.21)$$

$$\frac{\partial \phi_{k_2}}{\partial t} = -\frac{1}{B_0}\frac{\rho^2}{1 + k^2\rho^2}(\boldsymbol{k} \times \boldsymbol{k}_1) \cdot \hat{\boldsymbol{z}}(k_1^2 - k^2)\phi_k\phi_{k_1}^*\,\mathrm{e}^{-\mathrm{i}\Delta\omega t} \qquad (7.22)$$

where the mismatch is $\Delta\omega = \omega - \omega_1 - \omega_2$.

We can now substitute the matching condition for the wave vectors into equation (7.22) in order to eliminate $\boldsymbol{k}$. The vector products may then all be expressed in terms of $(\boldsymbol{k}_1 \times \boldsymbol{k}_2) \cdot \hat{\boldsymbol{z}}$. This leads, however, to a change of sign in all three equations. This means that the coupling factor of equation (7.21) will have opposite sign to the coupling factors of the other two equations. In this situation mode 1, i.e., the mode with the immediate magnitude of k, will act as a pump wave and we have cascading of wave quanta towards smaller and larger k according to figure 7.1. The threshold for the parametric interaction is

$$|\phi_1|^2 > \frac{\Delta\omega^2}{4V_0V_2} \qquad (7.23)$$

where V_0 and V_2 are the coupling factors. By using the dispersion relation in the form $k^2\rho^2 = k_y v_{De}/\omega_k - 1$, it is possible to write

$$\begin{aligned} k^2 - k_1^2 &= \omega_2 M \\ k_1^2 - k_2^2 &= = -\omega M \\ k_2^2 - k^2 &= \omega_1 M \end{aligned} \qquad (7.24)$$

where

$$M = \frac{v_{*e}}{3\rho^2\omega\omega_1\omega_2}[\omega(k_{y2} - k_{y1}) + \omega_1(k_y + k_{y2}) - \omega_2(k_y + k_{y1})].$$

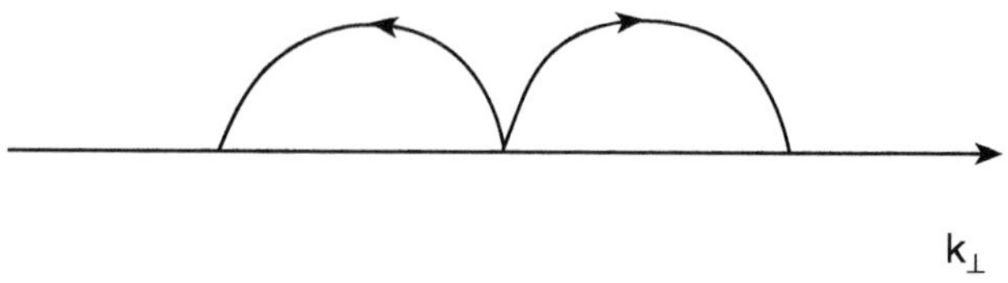

Figure 7.1. Double cascade.

We then find that when $k^2\rho^2 \ll 1$, i.e., $\Delta\omega \to 0$, the pump wave will be the mode with the largest frequency.

Assuming the presence of a large amplitude, long wavelength mode, Hasegawa and Mima derived, for a random phase situation, a stationary spectrum of the form

$$\left|\frac{e\phi_k}{T_e}\right|^2 = \Gamma\frac{(k\rho)^\alpha}{(1+k^2\rho^2)^\beta} \tag{7.25}$$

where $\alpha = 1.8$ and $\beta = 2.2$. This spectrum is in reasonable agreement with spectra observed in tokamak experiments where $10^{-6} < \Gamma < 10^{-5}$. Computer investigations by Fyfe and Montgomery [7.14] show a spectrum with the dependence $k^{-14/3}$ below and k^{-6} above a source, while recent experiments give the variation $k^{-3.5}$. The experiments, however, include several effects not included in equation (7.25), such as ion temperature effects and linear damping or growth. Another important phenomenon observed in nonlinear simulations of equation (7.11) is the generation of zonal flows [7.19]. Such flows may cause a stabilization of drift wave turbulence, leading to internal transport barriers.

For the system described by equations (7.15) and (7.16), the derivation of the coupling factors is considerably more complicated. The result may be written in the form

$$\frac{\partial\phi_k}{\partial t} = V_{12}\phi_{k_1}\phi_{k_2}\,\mathrm{e}^{\mathrm{i}\Delta\omega t} \tag{7.26}$$

where

$$V_{12} = \rho^2\frac{(\boldsymbol{k}_1\times\boldsymbol{k}_2)\cdot\hat{\boldsymbol{z}}}{k^2}\left[k_2^2 - k_1^2 - \rho^4k^2k_1^2k_2^2\frac{\omega\omega_1\omega_2}{\omega'\omega_1'\omega_2'}\left(\frac{v_g}{v_{*e}}\right)^2 \times\left(\frac{\omega_2}{\omega_2'}-\frac{\omega_1}{\omega_1'}\right)\right]\frac{\omega-\omega'}{2\omega-\omega'} \tag{7.27}$$

where $\omega' = k_yv_g$. The last factor gives the sign of the energy. The system described by equation (7.27) has the same cascade rules, determined by k numbers, as equations (7.20)–(7.22). There is, however, no transition to the usual weak turbulence rule for perfect matching. The wave energy is given by

$$W_k = k^2\rho^2\left(\frac{e\phi_k}{T_e}\right)^2 + \frac{v_g}{v_{*e}}\left(\frac{n_k}{n_0}\right)^2 = \frac{2\omega-\omega'}{\omega-\omega'}k^2\rho^2\left(\frac{e\phi_k}{T_e}\right)^2.$$

This wave energy is contrary to the usual weak turbulence case conserved also in the presence of mismatch. We also note that this expression for the wave energy, obtained from a nonlinear conservation relation in a fluid model, agrees with the expression in equation (4.79) obtained from a linear kinetic theory, to first order in the FLR parameter.

From the ion vortex equation we may derive a simple condition for the applicability of quasi-neutrality. Using the Poisson equation in equation (7.7) we have

$$\frac{\mathrm{d}}{\mathrm{d}t}\ln\left(\frac{\Omega_i + \Omega_{ci}}{n_e - \varepsilon_0 \Delta\phi/e}\right) = 0. \tag{7.28}$$

Since here n_e is the total electron density, we can expand the denominator for $n_e \gg \varepsilon_0 \Delta\phi/e$. Using equation (7.28) for Ω_i and assuming $\Omega_{ci} \gg \Omega_i$ we obtain, dropping $\Omega_i \varepsilon_0 \Delta\phi/e$

$$\frac{\mathrm{d}}{\mathrm{d}t}\ln\left[\left(\frac{1}{B_0}\Delta\phi\left(1 + \frac{\Omega_{ci}^2}{\omega_{pi}^2}\right) + \Omega_{ci}\right)\Big/ n_e\right] = 0. \tag{7.29}$$

We thus find the condition $\omega_{pi}^2 \gg \Omega_{ci}^2$ for quasi-neutrality. For a tokamak plasma we have typically $\omega_{pi} \sim 40\Omega_{ci}$, so the condition for quasi-neutrality is well fulfilled.

7.1.1 The nonlinear dielectric

An alternative to the previous formulation of the nonlinear dynamics in terms of the ion vortex equation is the formulation in terms of a dielectric function. For electrostatic modes this is

$$\omega \varepsilon_n(\omega, k) E_{\omega,k} = -\frac{1}{\varepsilon_0}\,\mathrm{i} J_{\omega,k}^2 \tag{7.30}$$

where $\varepsilon(\omega, k)$ is the linear dielectric function given by equation (4.66). For $k^2\lambda_{De}^2 \ll 1$ and $\tau \ll T_e$ we obtain

$$\varepsilon(\omega, k) = \frac{1}{k^2\lambda_{De}^2}\left(1 + k^2\rho^2 - \frac{\omega_{*e}}{\omega}\right). \tag{7.31}$$

The current $J^{(2)}$ is the nonlinear current. For electrostatic drift waves

$$J_{\omega,k}^{(2)} = en\boldsymbol{v}_{pi}^{(2)} = -en\frac{1}{B\Omega_{ci}}(\hat{\boldsymbol{z}} \times \nabla\phi) \cdot \nabla\nabla\phi \tag{7.32}$$

i.e., the nonlinear part of the polarization drift, neglecting v_{*i} when $T_i \ll T_e$. Substitution into equation (7.30) leads to

$$[\omega(1 - \rho^2\Delta) - \mathrm{i}\boldsymbol{v}_{*e} \cdot \nabla]\nabla\phi = -\mathrm{i}\Delta\frac{T_e}{eB\Omega_{ci}}(\hat{\boldsymbol{z}} \times \nabla\phi) \cdot \nabla\nabla\phi. \tag{7.33}$$

Taking the divergence of equation (7.33) we obtain

$$[\omega(1-\rho^2\Delta) - \mathrm{i}v_{*e}\cdot\nabla]\Delta\phi = -\mathrm{i}\rho^2\Delta(\hat{z}\times\nabla\phi)\cdot\nabla\Delta\phi.$$

Then inverting the Laplacian and transforming $\omega \rightarrow \mathrm{i}(\mathrm{d}/\mathrm{d}t)$, we obtain the Hasegawa–Mima equation

$$\frac{\mathrm{d}}{\mathrm{d}t}(\phi - \rho^2\Delta\phi) + \boldsymbol{v}_{*e}\cdot\nabla\phi = \rho^2(\hat{z}\times\nabla\phi)\cdot\nabla\Delta\phi \tag{7.34}$$

which is equivalent to equation (7.11).

We note that the particularly simple frequency dependence of equation (7.31) made it possible to transform to the time domain without expanding $\varepsilon(\omega, k)$ around a linear eigenfrequency. Because of this equation (7.34) is valid in the strongly nonlinear regime.

7.2 Diffusion

The main reason for the interest in collective perturbations in magnetized plasmas is the anomalous transport caused by a turbulence of such perturbations. The low frequency vortex modes treated here are of special interest for several reasons. First, we observe that a convection across the magnetic field is associated with the vorticity. Second, as we shall see in this section, low frequency modes cause efficient transport. Third, these modes are frequently driven unstable by inhomogeneities in pressure and magnetic field, making them hard to avoid in a confined plasma. Although the anomalous transport is of convective type it is usually treated as a diffusive process. This can be justified in a turbulent state, where the particle motion in the wave fields is stochastic and the requirement on particle stochasticity is in fact more easily fulfilled than the random phase approximation for the waves. For a stochastic motion of particles the diffusion coefficient is usually defined as

$$D = \lim_{t\rightarrow\infty}\frac{1}{2t}\langle\Delta r^2(t)\rangle \tag{7.35}$$

where Δr is the distance from the point where the particle was at $t = 0$, and $\langle\rangle$ denotes an average over all possible initial velocities or more generally n ensemble average. We now introduce the velocity $v(t)$ so that

$$\Delta r(t) = \int_0^t v(t')\,\mathrm{d}t'$$

and

$$\begin{aligned} D &= \lim_{t\rightarrow\infty}\frac{1}{2t}\left\langle\int_0^t \mathrm{d}t'\int_0^t \mathrm{d}t'' v(t')v(t'')\right\rangle \\ &= \lim_{t\rightarrow\infty}\frac{1}{2t}\int_0^t \mathrm{d}t'\int_0^t \mathrm{d}t''\langle v(t')v(t'')\rangle. \end{aligned}$$

We shall now assume that we have a stationary stochastic process such that $\langle v(t')v(t'')\rangle = \langle v(t'-t'')v(0)\rangle$. This means that the correlation between the velocities only depends on the difference in time $\tau = t' - t''$, and

$$D = \lim_{t\to\infty} \frac{1}{2t}\int_0^t \mathrm{d}\tau \int_0^t \mathrm{d}t'' \langle v(\tau)v(0)\rangle$$

which simplifies to

$$D = \int_0^\infty \langle v(\tau)v(0)\rangle \,\mathrm{d}\tau. \tag{7.36}$$

Another usual way of defining D is as the coefficient in the diffusion equation

$$\frac{\partial n}{\partial t} = D\frac{\partial^2 n}{\partial r^2} \tag{7.37}$$

where $n = n(r, t)$ is the particle density. A solution to equation (7.37) corresponding to the initial state where all particles are collected at $r = 0$ is

$$\tilde{n}(r, t) = \frac{N}{(4\pi Dt)^{1/2}}\,\mathrm{e}^{-r^2/4Dt} \tag{7.38}$$

where N is the total number of particles. Clearly, the probability of finding a particle between r and $r + \Delta r$ at time t is $\tilde{n}(r, t)\Delta r/N$ if Δr is small enough. This means that the ensemble average of a quantity $Q(r, t)$ can be written

$$\langle Q\rangle = \frac{1}{N}\int_{-\infty}^{\infty} \tilde{n}(r, t)Q(r, t)\,\mathrm{d}r. \tag{7.39}$$

As is easily seen we now have

$$\langle r^2\rangle = 2Dt$$

Since r here is the total deviation in position since $t = 0$, we realize that the two definitions of D are equivalent. In order to derive a useful expression for D for a time- and space-dependent process, it is convenient to start from equation (7.36), where $v(\tau)$ is represented in Fourier form

$$v(\tau) = v(r(\tau), \tau) = \frac{1}{(2\pi)^3}\int v_{k\omega}\,\mathrm{e}^{\mathrm{i}[\omega\tau - \boldsymbol{k}\cdot\boldsymbol{r}(\tau)]}\,\mathrm{d}\omega\,\mathrm{d}\boldsymbol{k}$$

and two space dimensions were assumed.

We then obtain from equation (7.36)

$$D = \int_0^\infty \mathrm{d}\tau \frac{1}{(2\pi)^3}\int \langle |v_{k\omega}^2|\rangle\,\mathrm{e}^{\mathrm{i}\omega\tau}\,\langle \mathrm{e}^{-\mathrm{i}\boldsymbol{k}\cdot\boldsymbol{r}(\tau)}\rangle\,\mathrm{d}\omega\,\mathrm{d}\boldsymbol{k} \tag{7.40}$$

where we assumed that $v_{k\omega}$ is uncorrelated with the phase function. In order to obtain D, we now need to know the velocity spectrum and the ensemble average of the space phase function. The latter can be obtained by using the representation in equation (7.39) of the ensemble average. This leads to the result

$$\langle \mathrm{e}^{-\mathrm{i}\boldsymbol{k}\cdot\boldsymbol{r}} \rangle = \mathrm{e}^{-k^2 D t}. \tag{7.41}$$

This result was verified numerically for thermal equilibrium by Joyce, Montgomery and Emery [7.10]. The characteristic time $(k^2 D)^{-1}$ is usually called the orbit decorrelation time, and is the time after which an average particle has moved so far due to diffusion that the field is uncorrelated with the field at the initial point. Specializing now to resonant modes, where $v_{k\omega} = v_k \delta[\omega - \omega(k)]$ and $\omega(k)$ is the solution of a dispersion relation, we find

$$D = \int_0^\infty \mathrm{d}\tau \frac{1}{(2\pi)^2} \int \langle v_k^2 \rangle \, \mathrm{e}^{\mathrm{i}\omega(k)\tau - k^2 D\tau} \, \mathrm{d}\boldsymbol{k} = \frac{1}{(2\pi)^2} \int \frac{\langle v_k^2 \rangle}{-\mathrm{i}\omega(k) + k^2 D} \, \mathrm{d}\boldsymbol{k} \tag{7.42}$$

where we assumed convergence at $\tau = \infty$, i.e., $\mathrm{Im}\,\omega < k^2 D$. Introducing now

$$\omega(k) = \omega_{kr} + \mathrm{i}\gamma_k$$

and the reality condition $\omega_{kr} = -\omega_{kr}$, we obtain

$$D = \frac{1}{(2\pi)^2} \int \langle v_k^2 \rangle \frac{k^2 D + \gamma_k}{\omega_{kr}^2 + (k^2 D + \gamma_k)^2} \, \mathrm{d}\boldsymbol{k}. \tag{7.43}$$

Equation (7.43) shows that the orbit decorrelation and the wave growth both contribute to the transport, while the real part of the eigenfrequency decreases the transport. For low frequency modes the dominant convective velocity is the $E \times B$ drift velocity. We then have

$$\boldsymbol{v}_k = \frac{\mathrm{i}}{B_0} (\hat{\boldsymbol{z}} \times \boldsymbol{k}) \phi_k.$$

The most efficient mode in a plasma in a homogeneous magnetic field is the convective cell mode. For this mode, $\omega_{kr} = 0$ and the orbit decorrelation usually dominates the damping. In this case we can solve equation (7.43) for D with the result

$$D = \frac{1}{B_0} \left(\int \frac{1}{2\pi} |\phi_k|^2 \mathrm{d}\boldsymbol{k} \right)^{1/2} \tag{7.44}$$

which is the diffusion coefficient for convective cells. It was first derived by Taylor and McNamara [7.7]. For a thermal equilibrium spectrum in the two-dimensional case

$$\frac{k^2 |\phi_k|^2}{8\pi} = \frac{T}{2\varepsilon} \tag{7.45}$$

where ε is the dielectric function. We thus obtain

$$D = \frac{1}{B_0} \left(\frac{2T}{\varepsilon} \ln \frac{L k_{\max}}{2\pi} \right)^{1/2} \tag{7.46}$$

where L is the maximum allowed wavelength (system dimension). The influence of ε was introduced by Okuda and Dawson [7.8]. The dielectric constant used was (compare equation (4.64)).

$$\varepsilon = 1 + \frac{\omega_{pi}^2}{\Omega_{ci}^2} + \frac{\omega_{pe}^2}{\Omega_{ce}^2}$$

which leads to a Bohm-like diffusion ($D \sim 1/B$) for $\omega_{pi}^2/\Omega_{ci}^2 \ll 1$, and to a diffusion independent of B for $\omega_{pi}^2/\Omega_{ci}^2 \gg 1$. This diffusion is, in the plateau regime, comparable to the classical diffusion, but is much larger in the Bohm regime. Most fusion machines are supposed to work in the plateau regime but here the anomalous transport will also dominate in a turbulent state, where the excitation level will be much larger than that given by equation (7.45).

Another mode of considerable interest is the magnetostatic mode (see section 5.6.2). This mode is electromagnetic and causes mainly electron diffusion by perturbing the magnetic flux surfaces. The velocity in equation (7.43) is here given by

$$v_k = v_{\parallel} \frac{\delta \boldsymbol{B}_{\perp}}{B_0}$$

where $\delta \boldsymbol{B}_{\perp}$ is the perturbation of the magnetic field perpendicular to the background magnetic field and $v_{\parallel}$ is the thermal velocity. This process was studied by Chu, Chu and Ohkawa [7.19], where the diffusion coefficient

$$D = \frac{T}{B_0} \left[\frac{2}{m L_{\parallel}} \ln \left(\frac{L \omega_p}{2\pi} \right) \right]^{1/2} \tag{7.47}$$

was obtained for a thermal equilibrium. Here $L_{\parallel}$ is the system length parallel to the magnetic field and L is the dimension in the perpendicular direction. This diffusion coefficient has a Bohm-like T/B scaling. Since this is mainly an electron diffusion, charge separation effects will efficiently prevent it from leading to actual particle transport. It will, however, cause a thermal conductivity instead, and it has been suggested that processes of this kind could explain the anomalous thermal conductivity of tokamaks, which is about two orders of magnitude larger than the classical. In the derivations of the diffusion coefficients in equations (7.46) and (7.47), it was assumed that the real part of the eigenfrequency could be neglected. This is not always a realistic assumption. For the convective cell mode, curvature of the magnetic field lines can violate this assumption, while for the magnetostatic mode a density inhomogeneity is enough. For both modes, magnetic shear can limit the maximum perpendicular extension of the mode. In such situations nonlinear modes driven by the ponderomotive force may often be more dangerous.

7.2.1 Fokker–Planck transition probability

The use of the solution of the diffusion equation for calculating ensemble averages can be generalized to solutions of the Fokker–Planck equation for diffusion in phase space [7.76]. We consider solutions of the equation

$$\left(\frac{\partial}{\partial t}+\boldsymbol{v}\frac{\partial}{\partial \boldsymbol{r}}\right)W(X,X',t,t')=\frac{\partial}{\partial \boldsymbol{v}}\left[\beta\boldsymbol{v}+D^{v}(t)\frac{\partial}{\partial \boldsymbol{v}}\right]W(X,X',t,t') \quad (7.48)$$

where $X=(\boldsymbol{r},\boldsymbol{v})$ is the phase space coordinate, the diffusion coefficient in velocity space D^{v} is, in general, a tensor, and β is the friction coefficient. Equation (7.48) has solutions of the form

$$\begin{aligned}W(X,X',t,t')&=\frac{1}{8\pi^{3}\Delta^{3/2}}\,\mathrm{e}^{3\beta\tau}\\&\quad\times\exp\left[-\frac{1}{2\Delta}(a_{i,j}\delta\rho_i\delta\rho_j+2h_{i,j}\delta\rho_i\delta P_j+b_{ij}\delta P_i\delta P_j)\right]\end{aligned} \quad (7.49)$$

where

$$\begin{aligned}\Delta&=\frac{1}{3}[a_{ij}b_{ij}-h_{ij}h_{ji}]\\a_{ij}\equiv a_{ij}(t,t')&=\frac{2}{\beta^2}\int_{t'}^{t}\mathrm{d}s\,D^{v}_{ij}(s)\\b_{ij}\equiv b_{ij}(t,t')&=2\int_{t'}^{t}\mathrm{d}s\,D^{v}_{ij}(s)\,\mathrm{e}^{2\beta(s-t')}\\h_{ij}\equiv h_{ij}(t,t')&=-\frac{2}{\beta}\int_{t'}^{t}D^{v}_{ij}(s)\,\mathrm{e}^{\beta(s-t')}\\\delta\rho&=\boldsymbol{v}\,\mathrm{e}^{\beta\tau}-\boldsymbol{v}' \qquad \tau=t-t'\\\delta\boldsymbol{P}&=\boldsymbol{r}-\boldsymbol{r}'+\frac{\boldsymbol{v}-\boldsymbol{v}'}{\beta}.\end{aligned}$$

For one-dimensional diffusion and a time-independent diffusion coefficient we obtain

$$a=\frac{2}{\beta^2}D^{v}\tau \qquad b=\frac{1}{\beta}D^{v}(\mathrm{e}^{2\beta\tau}-1)$$

$$h=-\frac{2}{\beta^2}D^{v}(\mathrm{e}^{\beta\tau}-1) \qquad \Delta=ab-h^2$$

and

$$W(X,X',\tau)=\frac{\mathrm{e}^{\beta\tau}}{2\pi\sqrt{\Delta}}\,\mathrm{e}^{-\frac{1}{2\Delta}(a\delta\rho^2+2h\delta\rho\delta P+b\delta P^2)}. \quad (7.50)$$

We may use equation (7.50) as a weight function to derive ensemble averages. Some examples are:

$$\langle \Delta r \rangle = \frac{v}{\beta}(1 - \mathrm{e}^{-\beta\tau})$$

$$\langle \mathrm{e}^{-k\Delta r(v,t,\tau)} \rangle = \exp\left[\mathrm{i}\frac{kv}{\beta}(1 - \mathrm{e}^{-\beta\tau}) - \frac{k^2}{\beta^2}\int_0^\tau \mathrm{d}\xi\, D^v(t-\xi)(1 - \mathrm{e}^{-\beta(\tau-\xi)})^2\right].$$

In the stationary case we have

$$\langle \mathrm{e}^{-k\Delta r} \rangle = \exp\left(ikv\tau - \frac{k^2 D^v}{3}\tau^3\right) \qquad (\beta\tau \ll 1) \qquad (7.51)$$

and

$$\langle \mathrm{e}^{-\mathrm{i}k\Delta r} \rangle = \exp\left(\frac{ikv}{\beta} - k^2 D\tau\right) \qquad (\beta\tau \gg 1) \qquad (7.52)$$

where $D = D^v/\beta^2$ is the diffusion coefficient in ordinary space. We now note that for strong friction, equation (7.52) reproduces the same result, i.e., equation (7.41) as we obtained with the transition probability equation (7.38). For small friction, however, we obtain equation (7.51), which is the result of the Dupree–Weinstock renormalization [7.3–7.5].

We may now obtain the corresponding solution in the non-Markovian case as a convolution in time of equation (7.50). It can be rewritten in terms of Fourier components in time of $D^v(t, \tau)$ and $\beta(t, \tau)$. From this formulation the diffusion coefficient equation (3.67) for diffusion in real space emerges in a natural way [7.81]. The result obtained here is, however, more general since it includes the nonlinear frequency shift.

7.3 Discussion

In this chapter we have derived the general form of the ion vortex equation, which can be used to describe most types of vortex mode in plasmas, as well as in fluids. Here we used it to derive nonlinear equations for drift waves and interchange modes. For these types of mode we discussed the dual cascade towards shorter and longer space scales, typical of two-dimensional systems. The cascade towards longer space scales is particularly important for transport and we generally need some damping mechanism for long wavelengths to obtain a realistic level of the transport. This mechanism will most likely be sheared plasma flows generated nonlinearly, or by neutral beams or neoclassical effects. These flows may create an absorbing boundary condition for long wavelengths if sufficiently long wavelengths are included in the system, as discussed in section 5.10.5. We also note the discussion of conservation relations and the comparison between the expressions for the wave energy of interchange modes obtained here, and from the dielectric properties in chapter 4. The calculation of diffusion from particle orbit integrations is a complement to the quasi-linear calculations in chapter 3. We note the convenient use of the solution of the diffusion equation as a weight function (transition probability) for calculating ensemble averages. This method was later extended to the general Fokker–Planck equation for diffusion in velocity space. From this calculation the renormalization by Dupree and Weinstock was recovered.

References

[7.1] Kadomtsev B B 1965 *Plasma Turbulence* (New York: Academic)
[7.2] Rudakov L I 1965 *Sov. Phys.–JETP* **21** 917
[7.3] Dupree T H 1967 *Phys. Fluids* **10** 1049
[7.4] Dupree T H 1966 *Phys. Fluids* **9** 1773
[7.5] Weinstock J 1969 *Phys. Fluids* **12** 1045
[7.6] Kadomtsev B B and Pogutse O P 1970 *Reviews of Plasma Physics* vol 5, ed M A Leontovitch (New York: Consultants Bureau) p 249
[7.7] Taylor J B and McNamara B 1971 *Phys. Fluids* **14** 1492
[7.8] Okuda H and Dawson J M 1973 *Phys. Fluids* **16** 408
[7.9] Davidson R C 1972 *Methods in Nonlinear Plasma Theory* (New York: Academic)
[7.10] Joyce G, Montgomery D C and Emery F 1974 *Phys. Fluids* **17** 110
[7.11] Hasegawa A 1975 *Plasma Instabilities and Nonlinear Effects* (New York: Springer)
[7.12] Weiland J and Wilhelmsson H 1977 *Coherent Nonlinear Interaction of Waves in Plasmas* (Oxford: Pergamon)
[7.13] Sanuki H and Schmidt G 1977 *J. Phys. Soc. Japan* **42** 260
[7.14] Fyfe D and Montgomery D 1979 *Phys. Fluids* **22** 246
[7.15] Sagdeev R Z, Shapiro V D and Shevchenko V I 1978 *Sov. J. Plasma Phys.* **4** 306
[7.16] Cheng C Z and Okuda H 1978 *Nucl. Fusion* **18** 87
[7.17] Tang W M 1978 *Nucl. Fusion* **18** 1089
[7.18] Hasagawa A and Mima K 1978 *Phys. Fluids* **21** 87
[7.19] Chu C, Chu M S and Ohkawa T 1978 *Phys. Rev. Lett.* **41** 653
[7.20] Hasagawa A, Maclennan C G and Kodama Y 1979 *Phys. Fluids* **22** 2122
[7.21] Hassam A B and Kulsrud R 1979 *Phys. Fluids* **22** 2097
[7.22] Navratil G A and Post R S 1979 *Comment. Plasma Phys. Control. Fusion* **5** 29
[7.23] Nozaki K, Taniuti T and Watanabe K 1979 *J. Phys. Soc. Japan* **46** 991
[7.24] Weiland J and Sanuki H 1979 *Phys. Lett.* A **72** 23
[7.25] Pavlenko V P and Weiland J 1980 *Phys. Rev. Lett.* **44** 148
[7.26] Pavlenko V P and Weiland J 1980 *Phys. Fluids* **13** 408
[7.27] Weiland J 1980 *Phys. Rev. Lett.* **44** 1411
[7.28] Okuda H 1980 *Phys. Fluids* **23** 498
[7.29] Hasagawa A, Okuda H and Wakatani M 1980 *Phys. Rev. Lett.* **44** 248
[7.30] Weiland J, Sanuki H and Liu C S 1981 *Phys. Fluids* **24** 93
[7.31] Yu M Y, Shukla P K and Rahman H U 1981 *J. Plasma Phys.* **26** 359
[7.32] Shukla P K, Yu M Y, Rahman H U and Spatschek K H 1981 *Phys. Rev.* A **24** 1112
[7.33] Katou K 1981 *J. Phys. Soc. Japan* **51** 996

[7.34] Weiland J 1981 *Phys. Scr.* **23** 801

[7.35] Pavlenko V P and Weiland J 1981 *Phys. Rev. Lett.* **46** 246

[7.36] Nakach R, Pavlenko V P, Weiland J and Wilhelmsson H 1981 *Phys. Rev. Lett.* **46** 447

[7.37] Weiland J and Mondt J P 1982 *Phys. Rev. Lett.* **48** 23

[7.38] Rogister G and Hasselberg G 1982 *Phys. Rev. Lett.* **48** 249

[7.39] Taniuti T and Hasegawa A 1982 *Phys. Scr.* T **2:1** 147

[7.40] Bekki N, Takayasu H, Taniuti T and Yoshihara H 1982 *Phys. Scr.* T **2:2** 89

[7.41] Pecseli H 1982 *Phys. Scr.* T **2:1** 83

[7.42] Montgomery D C 1982 *Phys. Scr.* T **2:1** 83

[7.43] Hasagawa A and Wakatani M 1983 *Phys. Rev. Lett.* **50** 682

[7.44] Hasagawa A and Wakatani M 1983 *Phys. Fluids* **26** 2770

[7.45] Waltz R E 1983 *Phys. Fluids* **26** 169

[7.46] Weiland J and Wilhelmsson H 1983 *Phys. Scr.* **28** 217

[7.47] Rogister G and Hasselberg G 1983 *Phys. Fluids* **26** 1467

[7.48] Rahman H U and Weiland J 1983 *Phys. Rev.* A **28** 1673

[7.49] Terry P and Horton W 1983 *Phys. Fluids* **26** 106

[7.50] Hazeltine H D 1983 *Phys. Fluids* **26** 3242

[7.51] Pecseli H L, Mikkelsen T and Larsen S E 1983 *Plasma Phys.* **25** 1173

[7.52] Pecseli H L, Rasmussen J J, Sugai H and Thomsen K 1984 *Plasma Phys. Control. Fusion* **26** 1021

[7.53] Weiland J 1984 *Phys. Scr.* **29** 234

[7.54] Shukla P K, Yu M Y, Rahman H U and Spatschek K H 1984 Nonlinear convective motion in plasmas *Phys. Rep.* **105** 227

[7.55] Weiland J and Mondt J P 1985 *Phys. Fluids* **28** 1735

[7.56] Liewer P C 1985 *Nucl. Fusion* **25** 543

[7.57] Waltz R E 1985 *Phys. Lett.* **55** 1098

[7.58] Petviashvili V I and Pokhotelov O A 1985 *JETP Lett.* **42** 54

[7.59] Shukla P K 1985 *Phys. Rev.* A **32** 1858

[7.60] Witalis E A 1986 *IEEE Trans. Plasma Sci.* **14** 842

[7.61] Turner L 1986 *IEEE Trans. Plasma Sci.* **14** 849

[7.62] Hsu C T, Hazeltine H D and Morrison J P 1986 *Phys. Fluids* **29** 1480

[7.63] Taniuti T 1986 *J. Phys. Soc. Japan* **55** 4253

[7.64] Jovanovic D, Pecseli H L, Rasmussen J J and Thomsen K 1987 *J. Plasma Phys.* **37** 81

[7.65] Liljeström M and Weiland J 1988 *Phys. Fluids* **31** 2228

[7.66] Katou K and Weiland J 1988 *Phys. Fluids* **31** 2233

[7.67] Nordman H and Weiland J 1988 *Phys. Lett.* A **37** 4044

[7.68] Martins A M and Mendoca J T 1988 *Phys. Fluids* **31** 3286

[7.69] Shukla P K and Weiland J 1989 *Phys. Lett.* A **136** 59

[7.70] Shukla P K and Weiland J 1989 *Phys. Rev.* A **40** 341

[7.71] Nordman H and Weiland J 1989 *Nucl. Fusion* **29** 251

[7.72] Hong B G and Horton W 1989 *Phys. Fluids* B **2** 978

[7.73] Nordman H, Weiland J and Jarmén A 1990 *Nucl. Fusion* **30** 983

[7.74] Wilhelmsson H 1990 *Nucl. Phys.* A **518** 84

[7.75] Persson M and Nordman H 1991 *Phys. Rev. Lett.* **67** 3396

[7.76] Weiland J and Nordman H 1991 *Nucl. Fusion* **31** 390

[7.77] Nycander J and Yankov V V 1995 *Phys. Plasmas* **2** 2874

[7.78] Pecseli H L and Trulsen J 1995 *J. Plasma Phys.* **54** 401
[7.79] Mattor N and Parker S E 1997 *Phys. Rev. Lett.* **79** 3419
[7.80] Throumoulopoulos G N and Pfirsch D 1997 *Phys. Rev.* E **56** 5979
[7.81] Zagorodny A and Weiland J 1999 *Phys. Plasmas* **6** 2359

Answers to Exercises

2.1
$$\frac{\delta n}{n} \Big/ \frac{e\phi}{T}.$$

2.2 v_g equals twice the curvature drift after averaging over a Maxwellian distribution.

2.4 The diamagnetic drift is divergence free when ∇P is parallel to ∇n.

3.1 This is due to the fact that nv_* is divergence free, see equation (1.4).

3.2 (a) No difference.

3.2 (b) The only difference is that $k_y^2\rho^2$ is replaced by $k_\perp^2\rho^2$.

3.3 These are the effects giving the finite $\nabla \cdot \boldsymbol{A}$ (compare the discussion following equation (1.7)). This means that both kinds of ion inertia, appearing as $k_y^2\rho^2$ and $k_\parallel^2 c_s^2$, are associated with compressibility.

3.4 In both cases the inertia term $\omega(\omega - k_y v_{gi})$ is replaced by $\omega(\omega - \omega_{*i} - k_y v_{gi})$.

3.5 The solution of the dispersion relation can be written $\omega = \omega_r + \mathrm{i}\gamma$, where

$$\omega_r = \omega_{*e}(1 - k_y^2\rho^2) + k_y v_{gi}$$
$$\gamma = \frac{m_e \nu_{ei}}{k_\parallel^2 T_e}\omega_{*e}[k_y^2\rho^2\omega_{*e} + k_y(v_{ge} - v_{gi})].$$

3.6
$$\beta < -m^2 \Big/ \frac{\mathrm{d}\ln P}{\mathrm{d}\ln r}.$$

3.7
$$\beta_c = \left(\frac{a}{qR}\right)^2.$$

3.8 The intermediate result is

$$\frac{\delta n_e}{n} = \frac{\omega_{*e}}{\omega}\frac{e\phi}{T_e} - k^2\rho^2\frac{k_\parallel^2 v_A^2}{\omega^2}\frac{\omega}{k_\parallel c}\frac{eA_\parallel}{T_e}.$$

3.9

$$E_\parallel = -\mathrm{i}k_\parallel\phi\frac{k^2\rho^2 k_\parallel^2 v_A^2}{\omega(\omega_{*e} - \omega) + k^2\rho^2 k_\parallel^2 v_A^2 + k_\parallel^2 c_s^2}.$$

3.10

$$m \approx 280 \text{ for } q = 2.$$

3.11
$$\omega^2(1 + k_y^2\rho^2) - \omega\omega_{*e}(1 - \tfrac{1}{2}k_y^2\rho_i^2) - k_\parallel^2 c_s^2 = 0.$$

Appendices

Appendix 1. Typical Parameter Values for a Tokamak Plasma

We shall here give numerical values of some of the most important quantities associated with low frequency modes in a tokamak plasma. The basic machine performance is taken from JET. With a magnetic field of

$$B = 27.7 \times 10^3 \text{ G} = 2.77 \text{ T}$$

we find the cyclotron frequencies

$$\Omega_{ci} = 2.65 \times 10^8 \text{ s}^{-1}$$

$$\Omega_{ce} = 4.87 \times 10^{11} \text{ s}^{-1}.$$

A density of

$$n_0 = 10^{20} \text{ m}^{-3}$$

corresponds to the plasma frequencies

$$\omega_{pe} = 5.6 \times 10^{11} \text{ s}^{-1}$$

$$\omega_{pi} = 1.3 \times 10^{10} \text{ s}^{-1}$$

and the Alfvén velocity

$$v_A = B_0/(\mu_0 n_0 m_i)^{1/2} = 0.6 \times 10^7 \text{ m s}^{-1}.$$

This gives the dielectric constant for flute modes ($k_\parallel = 0$)

$$\varepsilon = 1 + \mu_0 \rho_m c^2/B_0^2 = 1 + c^2/v_A^2 = 1 + \frac{\omega_{pi}^2}{\omega_{ci}^2} \approx 1 + \left(\frac{\rho}{\lambda_{de}}\right)^2 \approx 2406 \left(\frac{\rho}{\lambda_{de}} = 49\right)$$

where ρ_m is the mass density and $\rho = c_s/\Omega_{ci}$.

We also notice that

$$\omega_{pe} = 1.15\Omega_{ce}.$$

At fusion temperatures

$$T_e = T_i = 10^8 \text{ K} = 8.6 \text{ keV} = 1.38 \times 10^{-15} \text{ J}.$$

We find

$$\begin{aligned}
\lambda_{De} &= (\epsilon_0 T_e/ne^2)^{1/2} = 0.69 \times 10^{-2} \text{ cm} \\
g &= (n\lambda_{De}^3)^{-1} = 3 \times 10^{-8} \\
v_{\text{th}\,e} &= (2T_e/m_e)^{1/2} = 0.55 \times 10^8 \text{ m s}^{-1} \\
v_{\text{th}\,i} &= 1.29 \times 10^6 \text{ m s}^{-1} \\
\rho_e &= v_{\text{th}\,e}/\Omega_{ce} = 1.13 \times 10^{-2} \text{ cm} \\
\rho_i &= 0.49 \text{ cm} \\
\nu_{ei} &= 0.5 \times 10^4 \text{ s}^{-1} \\
D_e &= \nu_{ei}\rho_e^2 = 0.6 \times 10^{-4} \text{ m}^2 \text{ s}^{-1}.
\end{aligned}$$

With a major radius

$$R = 3 \text{ m}$$

and a minor radius

$$a = 1 \text{ m}$$

we find

$$\begin{aligned}
\kappa &= 1/a = 1 \text{ m}^{-1} \\
v_{*e} &= \frac{\kappa T_e}{eB_0} = 3.2 \times 10^3 \text{ m s}^{-1} \\
v_g &= \frac{g}{\Omega_{ci}} = \frac{T_i/m_i}{r\Omega_{ci}} = 1.1 \times 10^3 \text{ m s}^{-1} \\
\omega_{int} &= (\kappa g)^{1/2} = 5 \times 10^5 \text{ s}^{-1}.
\end{aligned}$$

For $q = 2$ and $k_\parallel = 1/qR$, we have

$$k_\parallel v_A = 10^5 \text{ s}^{-1}$$

with a plasma current

$$I = 2.6 \times 10^6 \text{ A}$$

we have the average electron velocity

$$v_{be} = 0.5 \times 10^5 \text{ m s}^{-1}.$$

Appendix 2. Useful Vector Relations

$$\boldsymbol{A} \times (\boldsymbol{B} \times \boldsymbol{C}) = (\boldsymbol{A} \cdot \boldsymbol{C})\boldsymbol{B} - (\boldsymbol{A} \cdot \boldsymbol{B})\boldsymbol{C} \tag{A.1}$$

$$\nabla \cdot (\boldsymbol{A} \times \boldsymbol{B}) = \boldsymbol{B} \cdot (\nabla \times \boldsymbol{A}) - \boldsymbol{A} \cdot (\nabla \times \boldsymbol{B}) \tag{A.2}$$

$$\nabla \cdot (\phi \boldsymbol{A}) = \phi \nabla \cdot \boldsymbol{A} + \boldsymbol{A} \cdot \nabla \phi \tag{A.3}$$

$$\nabla \times (\phi \boldsymbol{A}) = \phi \nabla \times \boldsymbol{A} - \boldsymbol{A} \times \nabla \phi \tag{A.4}$$

$$\nabla \times (\boldsymbol{A} \times \boldsymbol{B}) = \boldsymbol{A}(\nabla \cdot \boldsymbol{B}) - \boldsymbol{B}(\nabla \cdot \boldsymbol{A}) + (\boldsymbol{B} \cdot \nabla)\boldsymbol{A} - (\boldsymbol{A} \cdot \nabla)\boldsymbol{B} \tag{A.5}$$

$$\nabla(\boldsymbol{A} \cdot \boldsymbol{B}) = \boldsymbol{A} \times (\nabla \times \boldsymbol{B}) + \boldsymbol{B} \times (\nabla \times \boldsymbol{A}) + (\boldsymbol{A} \cdot \nabla)\boldsymbol{B} + (\boldsymbol{B} \cdot \nabla)\boldsymbol{A} \tag{A.6}$$

$$\nabla \times (\nabla \times \boldsymbol{A}) = \nabla(\nabla \cdot \boldsymbol{A}) - \Delta \boldsymbol{A} \tag{A.7}$$

$$\nabla \cdot (\nabla \times \boldsymbol{A}) = 0 \tag{A.8}$$

$$\nabla \times \nabla \phi = 0 \tag{A.9}$$

$$(\boldsymbol{A} \times \boldsymbol{B}) \cdot (\boldsymbol{C} \times \boldsymbol{D}) = (\boldsymbol{A} \cdot \boldsymbol{C})(\boldsymbol{B} \cdot \boldsymbol{D}) - (\boldsymbol{A} \cdot \boldsymbol{D})(\boldsymbol{B} \cdot \boldsymbol{C}). \tag{A.10}$$

Appendix 3. Cylindrical Coordinates

$$\nabla^2 \phi = \frac{1}{r}\frac{\partial}{\partial r}\left(r\frac{\partial \phi}{\partial r}\right) + \frac{1}{r^2}\frac{\partial^2 \phi}{\partial \theta^2} + \frac{\partial^2 \phi}{\partial z^2}$$

$$\nabla \cdot \boldsymbol{A} = \frac{1}{r}\frac{\partial}{\partial r}(rA_r) + \frac{1}{r}\frac{\partial}{\partial \theta}A_\theta + \frac{\partial}{\partial z}A_z$$

$$\nabla \times \boldsymbol{A} = \left(\frac{1}{r}\frac{\partial A_z}{\partial \theta} - \frac{\partial A_\theta}{\partial z}\right)\hat{\boldsymbol{r}} + \left(\frac{\partial A_r}{\partial z} - \frac{\partial A_z}{\partial r}\right)\hat{\boldsymbol{\theta}} + \left[\frac{1}{r}\frac{\partial}{\partial r}(rA_\theta) - \frac{1}{r}\frac{\partial A_r}{\partial \theta}\right]\hat{\boldsymbol{z}}.$$

General References

The following references are particularly relevant to the general area covered by this book.

Plasma Physics for Magnetic Fusion

[1] Krall N A and Trivelpiece A W 1973 *Principles of Plasma Physics* (New York: McGraw-Hill)

[2] Goldston R J and Rutherford P H 1995 *An Introduction to Plasma Physics* (Bristol: Adam Hilger)

Theory of Stability and Transport in Magnetic Confinement Systems

[3] Manheimer W M and Lashmore Davies C N 1989 *MHD and Microinstabilities in Confined Plasma* (Bristol: Adam Hilger)

[4] Hasegawa A 1975 *Plasma Instabilities and Nonlinear Effects* (New York: Springer)

[5] Kadomtsev B B and Pogutse O P 1970 *Reviews of Plasma Physics* vol 5 (New York: Consultants Bureau) p 249

[6] Connor J W and Wilson H R 1994 *Plasma Phys. Control. Fusion* **36** 719

Nonlinear Effects and Turbulence

[7] Kadomtsev B B 1965 *Plasma Turbulence* (New York: Academic)

[8] Davidson R C 1972 *Methods in Nonlinear Plasma Theory* (New York: Academic)

[9] Weiland J and Wilhelmsson H 1977 *Coherent Nonlinear Interaction of Waves in Plasmas* (Oxford: Pergamon)

Theory and Experiments on Transport

[10] Liewer P C 1985 *Nucl. Fusion* **25** 543

[11] Wagner F and Stroth U 1994 *Plasma Phys. Control. Fusion* **36** 719

[12] Carreras B A 1997 *IEEE Trans. Plasma Sci.* **25** 1281

References [1] and [2] are comprehensive and include general plasma physics with applications to magnetic fusion. They treat difficult and fundamental problems rigorously and provide excellent basic knowledge for a fusion physicist.

References [3] and [4] are similar to the present book in that they treat both MHD and transport. References [4] discusses several instabilities in the context of space physics and also includes nonlinear effects.

References [5] and [6] are review papers that discuss many instabilities of interest for transport. Reference [6] also presents transport coefficients corresponding to many instabilities.

Reference [7] is the first, and also the most frequently cited, book on plasma turbulence. It is mainly focused towards problems relevant to magnetic fusion and also contains one of the first renormalizations of plasma turbulence.

Reference [8] is particularly strong on kinetic nonlinear theory. It includes several mathematical tools such as the method of multiple time scales.

Reference [9] is more directed towards general plasma physics and laser fusion. It does, however, cover problems of nonlinear dynamics and partially coherent wave interactions relevant to the nonlinear saturation of drift wave turbulence.

References [10] and [11] discuss experimental transport research, including diagnostics, in detail. Reference [12] is more focused on the relevance of different theories for explaining experimental results.

Index